Algebra and probability

for 6 to 11 year olds

Kid-Zebra-ability

Written by

Algebra and Probability for kids

M. Saiprasad.

B.Sc(Special Maths), B.E(civil), MIE(India)

Edited by

Pramod Pillari, Florida

Algebra strikes you everywhere. Without it your kid feels strained at times to understand simple subjects like geometry.

This book contains chapters pertaining to introduction of Algebra and probability concepts to kids.

Your kid can do few pages each week. Do not insist more.

mathematics

Contents

Algebra 12
    variables and constants examples (3 examples) 12
    Variables and constants Exercise (8 problems) 13
        1)   In (-21 x), ... is constant and .... is variable 13
        2)   In 20 x y, ____ is constant and ____ are variables 13
        3)   In 15mn/ k, ____ is constant and ____ are variables 13
        4)   In 7a/ 13b, ____ are constants and ____ are variables 13
    Four fundamental operations on variables 14
        Example 1: Add a and b; a, b are whole numbers 14
        Example 2: Add a + b and c; a, b are whole numbers 14
        Example 3: add the variables A + A + A 14
        Example 4: Add (x + y) and (y + z); x, y are whole numbers 15
        Example 5: subtract a from b; a, b are whole numbers 15
        Example 5: Multiply a and b; a, b are whole numbers 15
        Example 5: Multiply a and a; a is whole number 15
        Example 6: Multiply (ab) and x 16
        Example 7: Multiply (ab) and xy 16
        Example 8: Multiply (ab) and (ac) 16
        Example 9: Multiply (ab) and (ac) 17
        Example10: Divide a by b 17
        Example 11: Divide ab by b 17
        Example 12: Divide ab with y 18
        Example 13: Divide ab with xy 18
        Example 14: Divide ab with aK 19
        Example 15: Simplify $ab \times xa \times bc \times cy$ 19
        Example 16: Sum of two numbers is 16. One number is 5. What is the other number 19
        Exercise: Just guess the answers without using pen and paper (12 problems) 20
    Naughty Question 21
Variable Expressions 21
    Exercise: Check the correct name of expression (8 problems) 22
    (say true or False) Algebraic variables - Examples 23

Algebra and Probability for kids

Exercise: Write the answers inside the given parentheses. Write each as an algebraic expression and give the answer where ever it is possible. First one is done for you. (44 problems) 24

Dependent and independent variables 25

Our friends x and y 25

Exercise (6 problems) 26

Variable Equations: 27

Mathematicians are lazy guys 27

List of symbols 28

Exercise: Give suitable answer with algebraic expression.(3 problems) 28

Finding the variable expressions according to question given (15 problems objective type). 29

Forming the variable expressions according to data. 30

Answers: 11) (x – 3)   12) (x + 3) 13) A -10 31

In – out tables 31

Find the function rule if not given or fill up the table when function rule is given. 31

About equations 33

Basic rule of equations: 34

Example : if x – 2 = 10 then what is x + 24 34

Exercise   Check True or false (10 problems) 35

Transformation of equations 36

Example : x + 8 = 40 what is x? 36

Example: x - 34 = 12 what is x? 37

Example 12 x = 60 what is x? 37

Example: x/29 = 6 what is x 38

Evaluate variables in the given expressions 39

Exercise(13 problems) 39

Exponential forms 40

Example 1. Evaluate 5^4 40

Read this as minus seven square 41

Example 5: How do you write exponent 4; base A 41

Problem 1: 125 is 5 to the power of ? 42

Problem 2: 64 is 8 to the power of ? 42

Problem 3: 32 is 2 to the power of ? 42

Problem 4: 64 is 4 to the power of ? 42

Problem 5: 512 is 8 to the power of ? 42

4

mathematics

- Problem 6: : $x^2 = 49$ then what is x? 42
- Problem 7: 125 is 5 to the power of ? 42
- Problem 8: $x^3 = 216$, then what is x? 42
- Problem 9: $x^4 = 16$, then what is x? 42
- Problem 10: 64 is 4 to the power of ? 42
- Problem 11: 512 is 8 to the power of ? 42
- Problem 12: 243 is 7 to the power of ? 42

Multiplication of exponent expressions. 42
- Examples: multiply following expressions (3 examples) 43
- Example: Prove $x^0 = 1$ 43
- Exercise on finding exponential forms (6 examples) 44

Division of exponentials (4 examples) 44
- Example 1: x^7 / x^3 44
- Example 2: x^4 / x^2 45
- Example 3: $a^{10} b^8 c^{12} / a^6 b^3 c^7$ 45
- Example 4: $1 / x^3$ 45
- Exercise: Perform multiplication (11 problems) 45
- Exercise: Perform Division 46
- Exercise: Solve the variable expressions (16 problems) 46

Solving equations 47
- Exercise: Find the variable present in the given equation (100 problems) 47
- Exercise: Find the variable present in the given equation (73 problems) 48

Algebraic terms 48

Like and unlike terms in polynomials 49
- Example: Add 2x and 3y 49
- Example: $4x^5 + 3 x^3 + 4xy + x y^5 + 5 xy + x^5 + 7 x^3 + 10 x y^5$ 50

Monomials 50
- A Binomial 51
- A trinomial 51

Identify the polynomial-type for each 52

Degree of polynomial 52
- Solve each equation 53

Distributive property of addition 54

Adding polynomials of single variable and constants 55
- Exercise simplify the following expressions(5 problems) 55
- Exercise Solve for the variable (75 problems) 55

Algebra and Probability for kids

Exercise: Simplify each expression.(28 problems) 58

Exercise: Simplify each expression(18 problems) 59

Word problems 60

Problem1: Vihaan said to Ilo. Imagine a number and think so many dollars are there in your pocket 60

Problem 2: 62

A man wanted to go home and give flowers to his wife. He picked some flowers. Before reaching home he had to take bath in three water pools...................62

He got into first pool took his bath. When he came out he noticed that the flowers in his bag had doubled. He threw three flowers into pool as mark of gratitude..62

He went to second pool and took bath. When he came out he noticed that the number of flowers in the bag doubled again. "Thanks! He told to pool and dropped five flowers in the pool and carried on to the third and final pool. After taking bath in the third pool had again noticed that the flowers have doubled. He gave ten flowers to the pool and stepped into hid house but there were no flowers in the bag to give to his wife. 62

Can any one tell how many flowers he had first, in his bag. 62

Problem 3: Few birds were flying in the sky. And few birds are sitting on ground. A bird sitting on floor said to birds flying in the sky "If one of you come down and join us we both have same number of birds." Then the birds in the sky said to the birds sitting on ground " If one of you fly and join us we will be double in number to you sitting on ground" How many birds are flying and how many birds are sitting on the ground? 63

Problem 4: The sum of the digits in a two-digit number is 5. If we interchange the digits then the new number formed is 9 less. Find the original number. 64

Problem 5: There are 39 boys in the class. This is three more than four times the number of girls. How many girls are in the class?...65

Problem 5a: A rooster has laid eggs 10 times to the number of its legs. How many eggs did it lay?...65

Problem 6: The sum of two numbers is 40, and one of them is 14 more than the other. What are the two numbers?...66

Problem 7: Half a number plus 5 is 43.What is the number?..66

Problem 8: The sum of two consecutive integers is 35. What are the two numbers? 67

Problem 9: The sum of two consecutive even integers is 30. What are the two numbers?..67

Problem 11: six years ago, Victor's age was half of the age he will be in 10 years. How old is he now? 68

Problem 12: A bat and a ball cost one dollar and ten cents in total. The bat costs a dollar more than the ball. How much does the ball cost?..............................69

mathematics

Problem 13: The sum of two numbers is 51. One of the numbers exceeds the other by 9. Find the numbers. 70

Problem 14: The difference between the two numbers is 54. The ratio of the two numbers is 5: 4. What are the two numbers? 70

Problem 15: The length of a rectangle is twice its breadth. If the perimeter is 96 meter, find the length and breadth of the rectangle. 71

Problem 16: Vicky is 5 years younger than Ricky. Four years later, Ricky will be twice as old as Vicky. Find their present ages. 72

Binomials: The polynomial with two monomials is called by binomial. 73

Example: Add $2x^2y + 3a^2b$ and $x^2y - a^2b$ 74

Example: Subtract $2x - 4y$ from $5x + 8y$ 74

Example: Subtract $x^2y^3 - a^2b$ from $2x^2 y^3 + 3a^2b$ 75

Exercise (40 problems) Simplify each expression. 76

Solving linear equations with two variables 77

Example Solve the equations $2x + 3y = 7$ and $x + 2y = 8$ 78

Multiplication of binomials 79

Example: $7x^3 (5x + y^2)$ 79

Example: $-7a (4a - 6b)$ 80

Example: $(6y + 3)(6y - 4)$ 80

Example: Expand $(3p + 2q)^2$ 81

Example: Expand $(a + b)^2$ 81

Example: Expand $(a - b)^2$ 81

Example: Expand $(a + b)(a - b)$ 82

Exercise: Find product. (28 problems) 82

Example 1: There are two whole numbers. The sum of twice one number and three times another number is 23 and their product is 20. Find the numbers. 83

Example 2: A triangle has a perimeter of 65cm. If 2 of its sides are equal and the third side is 5cm more than the equal sides, what is the length of the third side? 84

Example 3: A rectangle is 4 times as long as it is wide. If the length is increased by 4 inches and the width is decreased by 1 inch, the area will be 60 square inches. What were the dimensions of the original rectangle? 85

Example 4: Orion's father is 5 times older than Orion and Orion is twice as old as his sister Rema. In two years time, the sum of their ages will be 58. How old is Orion now? 86

Example 5: In the given parallelogram ABCD, find CD. 86

Exercise in solving linear equations with two variables 87

What is probability? 92

Probability Key Terms 92

Experiment 92
Outcome 93
Sample Space 93
Notation of Probability 93
Worked out examples. 94

    Example1: What is probability of striking red color in the spinning wheel shown. 94

    Example 2: What is probability of striking Yellow color in the spinning wheel shown. 96

    Example 3: A Spinner has four equal sectors with orange, blue, red and green colors. What is probability of landing on blue color on the spinning wheel, when it is spun. 96

    Example 4: A Dice is drawn. What is the probability of getting an even number? 97

    Example 5: A Dice is drawn. What is the probability of getting an odd number? 98

    Example 6: I asked peter to choose any number between 1 to twenty. What is likely hood of Peter choosing 12? 99

    Example 7: I asked peter to choose any number between 1 to twenty. What is likely hood of Peter choosing any prime number? 99

    Example 8: There are 7 four-point stars 7 six point stars, 6 diamonds and 6 fivepoint stars in inside a box. If one thing is to be picked up with closed eyes from the box what is probability of a diamond to come up? 100

    Example 9: John tosses a coin 50 times. What is probability of getting heads. 102

    Example 10: Bonne is asked to pick a card from a pack of cards. What are the chances he picks a King. 102

    Solution: 102

    Pack of cards has 52 cards. 102

    There are 4 kings in it. 102

    Therefore chances are 4/ 52 = 1/ 13 .... Ans 102

Terminology 102
Experiment 103
Outcome: 103
Sample space 103
"Equally Likely" 103
Compliment event 104
Mutually exclusive 104
Sure – Event 105
Impossible – Event 105

mathematics

Example 11: Bonne is asked to pick a card from a pack of cards. What are the chances he picks a King. 105
Solution: 105
Pack of cards has 52 cards. 105
There are 4 kings in it. 105
Therefore chance of picking King. 105
P(King) = 4/ 52 = 1/ 13 105
P'(King) = 1 - 105
or. 105
Probability of not picking a King is = 1 - 105
.... Ans 105
Example 12: There are 3 blue and 2 red marbles in a bag. What is the probability of drawing a blue marble on the first and second draw? 105
Example 13: The probability of event is always between what values? 106
Ans: between 0 and 1. 106
Example 14: Events are not affected by previous Events are called ? 106
Example 15: A number is chosen at random from 1 to 10. Find the probability of not selecting a multiple of 2 or a multiple of 3. 106
Example 15: Find the probability of drawing a 8 of Spades 106
Example 16: A number is chosen at random from 1 to 10. Find the probability of not selecting a multiple of 3. 106
Example 17: What is the total number of possible outcomes when rolling a pair of dice? 107
Example 18: Find the probability of drawing a Heart from a pack of cards. 107
Example 19: Find thc probability of drawing cards K, Q, J, A. 107
Example 20: Find the probability of drawing a black card. 107
Example 21: Find the probability of rolling an odd prime number. 107
Example 22: When two dice are rolled, find the probability of rolling prime numbers on both dice. 107
Example23: Find the probability of drawing a Diamond. In pack of cards. 107
Example 24: Find the probability of drawing a 8 of Clubs 108
Example 25: List all possible outcomes from rolling a die. 108
Example 26: Find the probability of drawing red cards 5 through 9 108
Example 27: Find the probability of drawing a Joker when there is one joker in entire pack. 108
Example 28: Find the probability of rolling a 4 or greater when you roll a dice or Find the probability of rolling at least 4 when you roll a dice 108
Example 29: Find the probability of not rolling a sum of 5 108

Exercise 109

Algebra

There are two words to know before learning algebra: Variable and Constant.

Variable is something that we don't know the value of. Its value can change in an expression. For example, we don't know how many apples there are in a basket; then we can call the number of apples in the basket as x, for mathematical purposes. Any letter can be used to represent a variable. Usually, lower case letters are used to write the variables.

mathematics

Constant is usually a number, whose value doesn't change. For example, 4 is a constant. Its value is always 4, regardless of which expression it is used in.

Sam is adding 4 more apples to the basket of apples, the total number of apples in the basket can be expressed as x + 4. This is because, we didn't know how many apples were in the basket, we considered that the basket had x number of apples. In this expression, x can be any number like 2 or 5 or 20. What Sam added to the basket is always a constant; because, we know it as 4.

Let's identify the variables and constants in the below expressions:

variables and constants examples (3 examples)

Problem 1: (a – 2) is an expression

Ans: a is the variable   2 is the constant

Problem 2: (4p + 5) is an expression

Ans: p is the variable   4 & 5 are the constants

Problem 3: (x - 3) is an expression

   Ans: x is the variable   3 is the constant

What is variable?

Algebra and Probability for kids

Variables and constants Exercise (8 problems)

1) In (-21 x), ... is constant and .... is variable.

2) In 20 x y, ____ is constant and ____ are variables.

3) In 15mn/ k, ____ is constant and ____ are variables.

4) In 7a/ 13b, ____ are constants and ____ are variables

5) In (3 − x) which is variable and which is constant?

   ___ is the variable   ___ is the constant

6) In (y + 4) which is variable and which is constant?

   ___ is the variable___ is the constant

7) In (56 / x) which is variable and which is constant?

   ___ is the variable   ___ is the constant

8) In (7y - 4) which is variable and which is constant?

   ___ is the variable___ is the constant

The problems in algebra are the problems we see in every day life, but in a different way. There is always something you feel you miss in the problem.

You can do the problem by intuition. But we do not know the exact process how we did it. Algebra does it for you.

Four fundamental operations on variables

Your kid needs to know how to add and present variables.

Example 1: Add a and b; a, b are whole numbers

Solution:

On adding a with b we get (a + b).

mathematics

Don't forget the parentheses. Brackets say that it is a single word in algebra............. **Ans**

Example 2: Add a + b and c; a, b are whole numbers

Solution:

$(a + b) + c = a + b + c = (a + b + c)$

Don't forget the parentheses in $(a + b + c)$. Brackets say that it is a single word in algebra.................. **Ans.**

Example 3: add the variables A + A + A

Solution:

Now there are all A^s. They are like terms. So The answer is 3A.

3A stands for 3 times A

$= A + A + A$

Example 4: Add (x + y) and (y + z); x, y are whole numbers

Solution:

$(x + y) + (y + z) = x + y + y + z$
$= x + 2y + z = (x + 2y + z)$

Note that like terms add up like ordinary numbers .... **Ans**

Example 5: subtract a from b; a, b are whole numbers

Solution:

On subtracting a from b we get (b - a).

Don't forget the parentheses.

Brackets say that it is a single word in algebra............. **Ans**

Example 5: Multiply a and b; a, b are whole numbers

Solution:

$a \times b = ab$ or (ab)

For product parentheses are many times not necessary. It is implied ab is a single term or monomial............ **Ans**

Example 5: Multiply a and a; a is whole number

Solution:

$a \times a = a^2$ because we are multiplying like terms or same terms   .... **Ans**

Example 6: Multiply (ab) and x

Solution:

We know that $ab = a \times b$

$(ab) \times x = a \times b \times x$ or (abx)

or (axb) or (xab)

Associate property holds for multiplication. According to this property of multiplication, multiplication can be done in any order............ **Ans**

Example 7: Multiply (ab) and xy

Solution:

(ab) × xy = a × b × x × y or (abxy) or (yaxb)

Associate property holds for multiplication. .... **Ans**

Example 8: Multiply (ab) and (ac)

Solution:
(ab) × (ac) = a × b × a × c

= a^2 × b × c = (a^2 b c)

When we multiply like terms, we represent the product with exponents

Exponents are again explained more in detail elsewhere in this book. .... **Ans**

Example 9: Multiply (ab) and (ac)

Solution:

$(a^2 b) \times (b q^4) = a^2 \times b \times b \times q^4$

$= a^2 \times b^2 \times q = (a^2 b q^4)$

$(a^2 b q^4)$ is one single term called monomial consisting of several terms again monomials. Product of several monomials is a monomial.   .... **Ans**

Example10: Divide a by b

Solution:

On dividing 'a' with 'b' we get quotient as $\dfrac{a}{b}$

.... Ans

Example 11: Divide ab by b

Solution:

On dividing ab with b we get quotient as $\dfrac{ab}{b}$

b and b cancel out and we get answer = a ... **Ans**

Example 12: Divide $\dfrac{a}{b}$ with y

Solution:

$\dfrac{a}{b} \div y = \dfrac{a}{b} \div \dfrac{y}{1}$

$$= \frac{a}{b} \times \frac{1}{y} = \frac{a}{by}$$ .... Ans

Example 13: *Divide* $\frac{a}{b}$ *with* $\frac{x}{y}$

Solution:

$$\frac{a}{b} \div \frac{x}{y} = \frac{a}{b} \times \frac{y}{x}$$

$$= \frac{ay}{bx}$$ .... Ans

Example 14: *Divide* $\frac{a}{b}$ *with* $\frac{a}{K}$

Solution:

$$\frac{a}{b} \div \frac{a}{K} = \frac{a}{b} \times \frac{K}{a}$$

$$= \frac{K}{b}$$ .... Ans

Example 15: *Simplify* $\frac{a}{b} \times \frac{x}{a} \times \frac{b}{c} \times \frac{c}{y}$

Solution:

a, b c cancel out

$$\frac{a}{b} \times \frac{x}{a} \times \frac{b}{c} \times \frac{c}{y} = \frac{x}{y}$$

Example 16: Sum of two numbers is 16. One number is 5. What is the other number

The answer is 11.

But we do not know how we did it. Answer struck the mind at once. It is just a … way good guess.

Steps involved in doing this problem according to algebra are many, as under.
- One number is 16. let the other number be A
- According to data given we form the algebraic equation as under
  A + 5 = 16
- Now eliminate 5 from left hand side. This we can do by subtracting 5 from both sides.
  A + 5 – 5 = 16 – 5
- New transformed equation is got by simplifying above
  A = 11
- This is the equation that gives us answer.

mathematics

Your mind did all these calculations.

That is guessing power if yours. God made your brain a super computer.

You will learn more of it in coming pages.

Exercise: Just guess the answers without using pen and paper (12 problems)

1) Sum of two numbers is 23. If one of those numbers is 18 what is other number?
2) Sum of two numbers is 51. If one of those numbers is 23 what is other number?
3) Sum of two numbers is -4. If 5 is one number what is the other?
4) Sum of -7 and what number gives us +9
5) Sum of +6 and what number gives us -9
6) Sum of (-74) and which number gives us -89
7) Product of two numbers is 45. If one number is 15, what is other number.
8) Product of two numbers is 9. If one of those two numbers is -3 what is other number.
9) When you divide this number by 8 you get the quotient 7 and remainder zero. What is that number.
10) When you divide this number by 10 you get the quotient 5 and remainder zero. What is that number.
11) When you divide this number by 5 you get the quotient 7 and remainder 3. What is that number.

12) When you divide this number by 9 you get the quotient 3 and remainder 4. What is that number.

Naughty Question

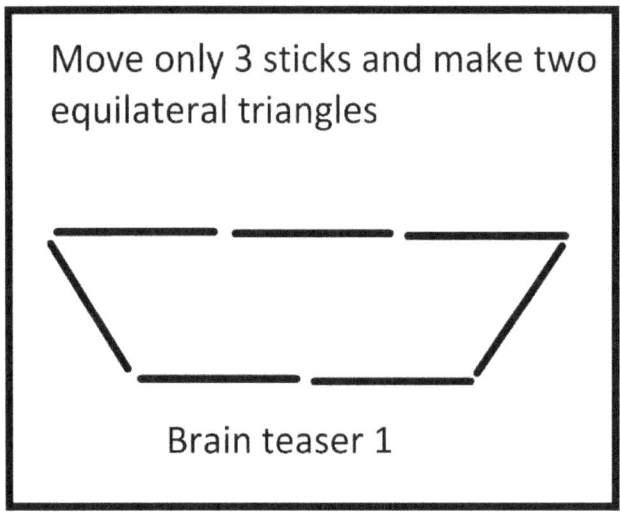

Variable Expressions

Expressions like a + b, 6x/5y are variable expressions. They involve variables like x, y.

These are also called Algebraic expressions. If the variable does not have negative powers, they are also called polynomials. Numbers 1, 7, 19, 456 like things are all numerical expressions. We know them as soon as we see them.

Exercise: Check the correct name of expression (8 problems)

1. 1+2 +3

    numerical expression   variable expression

2. 45 + 457 + 8763

    numerical expression   variable expression

3. 3/a + x^2

    numerical expression   variable expression

mathematics

4. v + u

   numerical expression   variable expression

5. 893

   numerical expression   variable expression

6. c

   numerical expression   variable expression

7. 3/2 + m²

   numerical expression   variable expression

8. g/h

   numerical expression   variable expression

In Algebra we do not know the value of the expressions. They are all unknown represented by variables like x and y.

When we do not know how many girls and how many boys are there in a class, we generally are not allowed to say what ever number that comes to our mind. If we do, we then estimate that erroneously. But in algebra we can.

Suppose there is class with girls and boys. We are free to say that there are A girls and B boys in that class.

Yes it is true.

To know what is 'A' and what is 'B' you either must have sufficient information or go to that school to see attendance register. It is just a matter of time.

Also you only said that there are 'A' number of boys and 'B' number of girls, If someone asks you what is the total strength of that class... you need not mince the words in algebra. It is A + B.

So, value of A + B depends upon values of both A and B

That's why we call (A+B) which is a single expression as 'dependent variable expression' while A and B are independent variable Expressions or simply independent variables.

(say true or False) Algebraic variables - Examples

a) The product of a and b = ab

b) The product of a and $\dfrac{b}{c} = \dfrac{ab}{c}$

c) If you divide x with y you get $\dfrac{x}{y}$

d) If you add p and q you get p + q

e) Subtracting 'e' from 'f' gives us (f - e)

I can join two pipes but not variables!

Ans: a) true b) True c) True d) True e) True

Exercise: Write the answers inside the given parentheses. Write each as an algebraic expression and give the answer where ever it is possible. First one is done for you. (44 problems)

1) u decreased by 17 ....[   ]

2) x increased by 6 ....[   ]

3) the product of x and 7 ....[   ]

4) the sum of q and 8 ....[   ]

mathematics

5) twice q if q = 6 ....[   ]

6) the quotient of 18 and n........[   ]

7) n cubed if n = 2........ [   ]

8) x^2 if x = 3 .... [   ]

9) a + 9 if a = 8

10) 19 − 3x, if x = 2

11) 5n -1- n^2, if n = 1

12) q - 2 if q = 8

13) x^2 if x = 3

14) 8 – p if p = 7

15) x + 8 if x = 9

16) n − 14 if n = 12

17) Product of 7 and k [   ]

18) one-fourth of x is added to the product of 7 and k

Answers:

1) u – 17, 2) x + 6, 3) 7x, 4) q+ 8, 5) 2q =12, 6) 18/n
7) 2^3 = 8, 8) 3^2 = 9, 9) 8+9 = 17
10) 19 – 6 = 13, 11) 5 - 1 - 1 = 3, 12) 6, 13) 9, 14) 1, 15) 17, 16) -2, 17) 7k, 18) (x/4) + 7k

Dependent and independent variables

In Algebra there are two kids of variables.

They are independent variables and dependent variables.

Consider this example.

A class has 30 students. If the number of boys who scored more than 50% marks in maths is 'x', the number of boys who secured less than 50% marks is "y", How is y related to x?

The answer obviously is y = (30 − x)

if x = 10, y = 20

if x = 5, y = 25

if x = 20, y = 10

Here x is independent variable and y is dependent variable.

y depends on x because y = (30 - x)

Our friends x and y

We use English letters to represent un-knowns.

From ages people have been using letter 'y' as their dependent variable and 'x' as their independent variable in the equations.

Let $y = x^2 + 2x + 7$

If x = 1, y = 10 and if x = 0, y = 7

As you change the value you give to x, the value of y keeps changing accordingly which is why we say y is dependent variable in that equation.

It should not be construed that one single letter like a, v, z stand always for independent variable. A variable expression like a + b + c, 4y, 6/d can also stand for dependent variable. This is because in algebra, (a + b + c), 4y, 6/d are single words or expressions

Exercise (6 problems)

Problem 1: On a tree there are K no of Pigeons and L number of parrots total of which is (K + L). Which is dependent variable here?

    1. K   2. L   3. (K + L)

mathematics

Problem 2: There are A no of pills and B number of tonics in a medical shop. Which is dependent variable here?

   1. B   2. A + B   3. A

Problem 3: There are V no of Mangoes and Z number of Avocados and K number of Pine apples in a basket. Which is dependent variable here?

   1. V   2. Z   3. (V + Z + K)

Problem 4: Write the dependent variable in the following equation $y = x^4 + x^2 + 2x + 5$

Problem 5: Write the dependent variable in the following equation $S = ut + ½ at^2$

Problem 6: Write the dependent variable in the following equation $E = mC^2$

Answers: (1) (k + L)   (2) (A + B)   (3) (V + Z + K) (4) y   (5) S   (6) E

Variable Equations:

If there is a statement comprising algebraic expressions to be are equal or not, such statement is called algebraic equation or variable equation.

You now know x + 6 is an algebraic expression.

Similarly a – 8 is also an algebraic expression.

Then we can write mathematical statements using these expressions and using the term " is equal to"

X +6 is equal to a – 8

Mathematicians are lazy guys

Mathematicians are lazy guys. They won't write full group of words like "Is equal to". Instead, they keep "=" sign and ask you to read it as "is equal to"

(1) $x + 6 = a - 8$ is an algebraic equation.

We read the statement as $(x+ 6)$ is equal to $(a – 8)$

(2) $x + 6 \neq a - 8$ is also an algebraic equation.

We read the statement as $(x+ 6)$ is not equal to $(a – 8)$

There are several symbols that are used in equations. We generally use is "="

In the future the kids will have to know every symbol. Why not introduce them now itself?

Curiosity sake let us collect some symbols that are used in Algebra.

List of symbols

Symbol	Read as
$=$	Is equal to
$\equiv$	Is identically equal to
$\approx$	Is approximately equal to
$\neq$	Is not equal to
$<$	Is less than
$\not<$	Is not less than
$>$	Is greater than
$\not>$	Is not greater than
$\leq$	Is less than or equal to
$\not\leq$	Is not less than or not equal to
$\not\geq$	Is not greater than or not equal to
$\Leftrightarrow$	That implies and implied by
$\Rightarrow$	That implies, that means

mathematics

Exercise: Give suitable answer with algebraic expression.(3 problems)

Problem 1: There are 100 students in a class. If the number of girls is x, and number of boys is 'y', then y = ?

   1. 100 × x   2. 100 + x   3. 100 - x

Problem 2: There are 230 animals in a Zoo. Number of horses = h, number is elephants = e, number of tigers is 't'. The remaining animals are y. Then, what is y?

   1. 230   2. 230 − (h+ e + t)   3. (h+ e + t)

Problem 3: John has x number of dollars in his house. He has money 'y' dollars in his bank which is 100 times of money he has in his house. What is y in terms of x?

   1. 100 × x   2. 100 + x   3. 100 ÷ x

Problem 4: what is 1/5th of the difference of a and 6.

Ans: (1) 100 − x   (2) 230 − (h+ e + t)   (3) 100 × x   (4) $\dfrac{a - 6}{5}$

Finding the variable expressions according to question given (15 problems objective type).

Problem 1: What is 5 more than XY?

   a. X + 5   b. Y + 5   c. XY + 5   d. XY + 2

Problem 2: What is 25 less than Y?

   a. K - 25   b. X - 25   c. Y - 25   d. A − 25

Problem 3: What is 340 less than Y?

   a. Y - 340   b. Y - 300   c. Y - 3405   d. Y − 1340

Problem 4: What is 500 more than X?

    a. 500 X   b. X + 5   c. X + 50   d. X + 500

Problem 5: What is 100 less than XY?

    a. X - 100   b. Y - 100   c. XY - 100   d. XY - 100

Problem 6: 6 divided by Y

    a. 6 + Y   b. Y - 6   c. Y ÷ 6   d. 6 ÷ Y

Problem 7: d is added to 9

    a. d - 6   b. d + 6   c. 6 d   d. 6/d

Problem 8: what is $1/6^{th}$ of x

    a. 6 times x   b. x – 6   c. $\frac{x}{6}$ d. x + 6

Forming the variable expressions according to data.

1) 3 divided by s   [   ]

2) d is added to 9   [   ]

3) Sum of 3 and h   [   ]

4) x divided by 4 [   ]

5) One-third of the sum of c and 6   [   ]

6) Three-fifths of c is added to 9

7) 5 is added to one-fifth of r   [   ]

8) Add one-sixth to 8 times y   [   ]

9) Subtract 8 from 9 times d   [   ]

10) 8 times the sum of 2 and h [   ]

Problem 11: Abby had 3 books more than Ashley. If Abby has X books how many books Ashley has?

    a) X + 3

    b) X - 3

    c) 3X

mathematics

   d) X ÷ 3

Problem 12: Abbie had 3 books more than Ashley. If Ashley has X books how many books Abbie has?

   a) X + 3

   b) X - 3

   c) 3X

   d) X ÷ 3

Problem 13: Mary had in her Garden 10 trees more than Pam' garden. If Mary had A trees how many trees Pam had?

   a) A times 10

   b) A + 10

   c) A - 10

   d) A ÷ 10

Answers: 11) (x – 3) 12) (x + 3) 13) A -10

In – out tables

Find the function rule if not given or fill up the table when function rule is given.

(1)

x	2x + 1
2	5
-6	
½	
3	

29

(2)

x	x³
2	8
-1	
½	
-2	

(3)

x	......
4	9
-6	1
½	5 ½
12	17

Function rule is, x becomes ....

(4)

1	4
2	5
3	6
¼	3 ¼
-3	0

mathematics

Function rule is, x becomes ....

Answers

(1)

x	2x + 1
2	5
-6	-11
½	2
3	7

(2)

x	x^3
2	8
-1	-1
½	1/8
-2	-8

(3) Function rule : x becomes x + 5

(4) Function rule is, x becomes x + 3

About equations

Equations are mathematical statements. They have LHS equated to RHS.

The statements are either true or false. They cannot both true and false. In mathematics if a statement is true then we say that the truth-value of the statement is "T". Otherwise we say the truth value of statement is "F"

When you manipulate equations either on LHS or on RHS, they lose their value

F becomes T and T may become F.

Here are some examples

$x - 2 = 10$ ....(T)

Add (-5) on RHS

$x - 2 = 10 - 5$ ....(F).

This new equation is no more true.

If manipulation is unavoidable, you should do it both sides. Add say 8 on both sides, the new equation will be

$X - 2 + 8 = 10 + 8$ .... (T)

That is because you can add same thing on both sides

You can not add one thing on one side and different thing on the other side.

If $x - 2 = 10$ .... (T)

$X - 2 + 8 = 10 + 6$ ....(F)

What was True hither to has become False now.

So make it sure that you add same thing on both sides.

This kind of manipulation on both sides of an equation is allowed under basic rules of equations.

Basic rule of equations:

The rule says

You can add same thing on both sides.

You can multiply with same thing on both sides

You can divide with same thing on both sides.

You can even subtract same thing on both sides.

Don't think manipulating equations is a silly thing. In-fact manipulation is unavoidable if you want to get required result.

mathematics

Example : if $x - 2 = 10$ then what is $x + 24$

Solution:

Suppose you know $x - 2 = 10$

But that is not what you want. You want to know what is

$x + 24$

Then you can do this.

Write the given equation.

$x - 2 = 10$

add 26 both sides

$\Leftrightarrow x - 2 + 26 = 10 + 26$

$\Leftrightarrow x + 24 = 36$ .... (1)

You got it. $x + 24 = 36$.

On left side what was $(x - 2)$ now is $x + 24$

On the right side what was 10 now became 36.

This process is called transformation of equations.

You can do the same thing in few more steps too duly eliminating things you don't require from the given expressions..

$x - 2 = 10$ we know it is true

$\Leftrightarrow x - 2 + 2 = 10 + 2$ .... (T)

$\Leftrightarrow x = 12$ (simplified) .... (T)

(add 24 to both sides of new transformed equation)

$\Leftrightarrow x + 24 = 12 + 24$ .... (T)

Simplify RHS of that equation you get

$x + 24 = 36$

You got the same answer as in (1).

Algebra and Probability for kids

Exercise   Check True or false (10 problems)

1) If $x = 6$ then $2x = 2 \times 6 = 12$ [T] [F]

2) If $a + b = c$ then $a + b + 6 = c + 5$ [T] [F]

3) If $c + d = K$ then $c + d + L = K + L$ [T] [F]

4) If $r = 2$ then $r \times 7 = 15$ [T] [F]

5) If $S = 100$ then $\dfrac{S}{2} = \dfrac{100}{2} = 50$ [T] [F]

6) If $x^2 + y^2 = Z$ then $x^2 + y^2 - 34 = Z - 34$ [T] [F]

7) If $ut = S - \dfrac{1}{2}at^2$ then $ut + \dfrac{1}{2}at^2 = S$ [T] [F]

8) If $u^2 + 2as = v^2$ then $u^2 + 2as + W = v^2 + W$ [T] [F]

9) If $c + d = K + d$ then $c = K - d$ [T] [F]

10) If $c = 2$ then $r \times C = 2r$ [T] [F]

Answers: 1) T 2) F 3) T 4) F 5) T 6) T 7) T 8) T 9) F 10) T

Transformation of equations

We can transform the equation without changing its truth value by using basic rule of equation. Basic rule says you can add or subtract, multiply or divide both sides of an equation by same number.

With this you can take the unwanted variable or numerical from one side of equation to the other side.

Example : x + 8 = 40 what is x?

Solution:

$x + 8 = 40$; truth value is (T) .... (1)

Here in this problem you have value of $x + 8$. But you need to know value of x but not $x + 8$.

Keeping basic rule in mind, let us eliminate that $+8$ which is glued to x on LHS.

mathematics

Subtract 8 from both sides.

Then our equation becomes $x + 8 - 8 = 40 - 8$

That is, $x = 40 - 8$ ....(2)

That is, $x = 32$

Notice that 8 which was on left side of (1) is now shifted to right side in the transformed equation (2), and also +8 changed into -8

The transformation has not changed truth value of given statement.

Example: x - 34 = 12 what is x?

Solution:

$x - 34 = 12$; truth value is (T) .... (1)

Here in this problem you have value of x - 34. But you need to know value of x but not x - 34.

Keeping basic rule in mind, let us eliminate that -34 which is glued to x in given equation $x - 34 = 12$.

To eliminate -34, add +34 on both sides.

Then our equation becomes

$x - 34 + 34 = 12 + 34$

$\Rightarrow x = 12 + 34$ ....(2)

$\Rightarrow x = 46$

Notice that 34 which was on left side of (1) is now shifted to right side in transformed equation (2), and also - 34 changed into +34.

The transformation has not changed truth value of given statement.

Example 12 x = 60 what is x?

Solution:

12 x = 60; truth value is (T) .... (1)

Here in this problem you have value of 12x. But you need to know value of x but not 12x.

Keeping basic rule in mind, let us eliminate that 12 which is glued to x on LHS.

Divide both sides by 12.

Then our equation becomes $\dfrac{12x}{12} = \dfrac{60}{12}$ ....(2)

That is, x = 5

Notice that 12 which was on left side of (1) is now shifted to right side in the transformed equation (2), and also multiplication with 12 changed into dividing with 12. Yet the transformation has not changed truth value of given statement.

Example: x/29 = 6 what is x

Solution:

$\dfrac{x}{29} = 6$ what is x?

$\dfrac{x}{29} = 6$; Its truth value is (T) .... (1)

Here in this problem you have value of $\dfrac{x}{29}$. But you need to know value of x but not $\dfrac{x}{29}$.

Keeping basic rule in mind, let us eliminate that 29 which is glued to x on LHS.

mathematics

Multiply both sides by 29.

Then our equation becomes

$$\frac{x}{29} \times 29 = 6 \times 29 \quad \ldots\ldots\ldots\ldots(2)$$

That is, $x = 174$

Notice that 29 which was on left side of (1) is now shifted to right side in the transformed equation (2), and also division with 29 changed into multiplying with 29. Yet the transformation has not changed truth value of given statement.

So generalizing the results, we can say

1) If $A + B = C$ then $A = C - B$
2) If $A - B = C$ then $A = C + B$
3) If $A \times B = C$ then $A = \frac{C}{B}$
4) If $\frac{A}{B} = C$ then $A = C \times B$

Evaluate variables in the given expressions

Exercise (13 problems)

Problem 1: What is q if $5q = 25$
    a. 3   b. 4   c. 2   d. 5

Problem 2: What is p if $\frac{p}{3} = 8$
    a. 10   b. 24   c. 12   d. 15

Problem 3: What is x^2 if $x + 6 = 9$
    a. 9   b. 4   c. 1   d. 3

Problem 4: What is x if $x - 7 = 10$
    a. 11   b. 13   c. 17   d. 15

Problem 5: What is k if 8 k = 16

    a) 3   b. 4   c. 2   d. 5

Problem 6: What is r if $\frac{r}{3} = 8$

    a. 10   b. 24   c. 12   d. 25

Problem 7: What is z if z + 4 = 10

    a. 6   b. 4   c. 1   d. 5

Problem 8: What is c if c − 6 = 6

    a. 10   b. 14   c. 12   d. 15

Problem 9: What is K if 5K = 20

    a. 10   b. 4   c. 12   d. 15

Problem 10: What is g if $\frac{g}{3} = 5$

    a. 10   b. 15   c. 12   d. 13

Problem 11: What is $4K^2 + k + 3$ if K = 2

    a. 10   b. 21   c. 12   d. 15

Problem 12: What is $y^3 + 120$ if y + 8 = 3

    a. -5   b. -4   c. -7   d. -6

Problem 13: What is d if d − 6 = 5

    a. 10   b. 11   c. 12   d. 10

    Answers: (1) d (2) b (3) a (4)c (5)c (6)b (7)a (8) c (9) b (10)b
    (11)16+2+3=21 (12)a (13)b

Exponential forms

Example 1. Evaluate 5^4

Solution:

In this expression 5 is called base and 4 is exponent.

Read this as '5 power 4' or "5 to the power of 4"

= 5 × 5 × 5 × 5

mathematics

= 625 → Multiply.

Example 2: Evaluate (-3)³.

Solution:

(-3)³ read this as minus 3 whole cube.

= (-3) × (-3) × (-3)

= -27

Example 3: evaluate -7².

Solution:

-7²

Read this as minus seven square.

= - (7²) = -(7 × 7) = -(49)

= - 49

Note: minus seven square ≠ minus seven whole square.

Example 4: Evaluate (2/5)³

Solution:

(2/5)³

= $\frac{2}{5} \times \frac{2}{5} \times \frac{2}{5}$

= $\frac{8}{125}$ .... Ans

Write each number as the power of a given base and given exponent.

Example 5: How do you write exponent 4; base A

Solution:

= A⁴

Example 6: Exponent 4; base is -A

Solution:

(-A)⁴ read as minus A whole power 4

Example 7: 125; write this as power of given base, base 5

Solution:

we know = (5) × (5) × (5) = 125

So, $125 = (5)^3$

Exercise(12 problems):

Problem 1: 125 is 5 to the power of ?.....................

Problem 2: 64 is 8 to the power of ?.....................

Problem 3: 32 is 2 to the power of ?.....................

Problem 4: 64 is 4 to the power of ?.....................

Problem 5: 512 is 8 to the power of ?.....................

Problem 6: : $x^2 = 49$ then what is x?

Problem 7: 125 is 5 to the power of ?.....................

Problem 8: $x^3 = 216$, then what is x?

Problem 9: $x^4 = 16$, then what is x?

Problem 10: 64 is 4 to the power of ?.....................

Problem 11: 512 is 8 to the power of ?.....................

Problem 12: 243 is 7 to the power of ?.....................

Answers: (1) 5^3 (2) 8^2 (3) 5 (4) 3 (5) 3 (6) 7 (7) 3 (8) 6 (9) 2 (10) 3 (11) 3 (12) 3

Multiplication of exponent expressions.

When we multiply one number with same number we can show it in exponential form.

Case1:

all like terms

$A \times A = A^2$

$A \times A \times A \times A \times A = A^5$ and so on.

Case2:

mathematics

Like and unlike terms

x × x × x × y × y × x × x × y × y × x

= x × x × x × x × x × x × y × y × y × y

= $x^4 y^4$

Case3:

Multiplying exponents

$x^4 \times x^5$ = (x × x × x × x) × (x × x × x × x × x)

= (x × x × x × x × x × x × x × x × x) = x^9

Notice that exponents added up after multiplication.

Examples: multiply following expressions (3 examples)

1. $x^2 \times x^6 = x^{2+6} = x^8$
2. $x^6 \times x^7 = x^{6+7} = x^{13}$
3. $x^4 \times y^5 \times x^4 \times y^4 \times x^5$
   $= x^{4+4+5} \times y^{5+4} = x^{13} y^9$

Example: Prove $x^0 = 1$

Solution:

We use transformation of equations

We can say $x^2 = x^2$, right?

That means $x^{2+0} = x^2$

That again means $x^2 \times x^0 = x^2$

Divide both sides of equation with x^2, then we have

$$\frac{x^2 x^0}{x^2} = \frac{x^2}{x^2}$$

That gives us $x^0 = 1$

Algebra and Probability for kids

Exercise on finding exponential forms (6 examples)

1) The correct exponential form of

   2(x)(x)(y)(y)(y) is ?

   $2x^3 y^2$ $2 x^3 y^3$ $2 x^2 y^3$ $2 x^2 y^2$

2) The correct exponential form of

   (m)(m)(m)(m)(n) is ?

   $m^3 n^2$ $m^4 n$ $m^2 n^3$ $m^2 n^2$

3) The correct exponential form of

   5(a)(b)(a)(b)(a) is ?

   $5 a^3 b^2$ $5 a^3 b^3$ $5 a^2 b^3$ $5 a^2 b^2$

4) The correct exponential form of

   4 (x)(x)(y)(y)(y) (x)(x)(y)(y)(y) is ?

   $4 x^5 y^2$ $4 x^4 y^3$ $4 x^2 y^5$ $4 x^4 y^5$

5) The correct exponential form of

   $(m)^2 (m) (m)^3 (m) (n) (n)^3$ is ?

   $m^8 n^4$ $m^7 n^7$ $m^7 n^4$ $m^8 n^8$

6) The correct exponential form of

   (p) (p)(p) $(p)^2$ is ?

   p^3 p^5 p^2 p^6

Division of exponentials (4 examples)

Example 1: x^7 / x^3

mathematics

$$\frac{x^7}{x^3} = \frac{x \times x \times x \times x \times x \times x \times x}{x \times x \times x}$$

$$= x \times x \times x \times x$$

$$= x^4 = x^{7-3}$$

Notice the exponents are subtracted

Example 2: x^4 / x^2

$$\frac{x^4}{x^2} = x^{4-2} = x^2$$

$$\frac{x^{13}}{x^4} = x^{13-4} = x^9$$

Example 3: $a^{10} b^8 c^{12} / a^6 b^3 c^7$

$$\frac{a^{10} b^8 c^{12}}{a^6 b^3 c^7} = a^{10-6} b^{8-3} c^{12-7}$$

$$= a^4 b^5 c^5$$

Example 4: $1 / x^3$

$$\frac{1}{x^3} = \frac{x^0}{x^3} = x^{0-3} = x^{-3}$$

Exercise: Perform multiplication (11 problems)

1) $n^3 \times n^2$
2) $2 s^{-6} b^4 \times s^3 b^{-2}$
3) $4y^3 \times 9 y^2$

Algebra and Probability for kids

4) $5K^{-2} \times 8K^3$

5) $4c^5 \times c^4$

6) $2g\,b^{-4} \times 9g^{-2}b^3$

7) $7r^6\,k^4 \times 2r^5\,k^3$

8) $4d\,g^{-6} \times 7d^{-2}\,g$

Answers: (1) n^5 (2) $3s^{-3}b^2$ (3) $36y^5$ (4) $40K$ (5) $4c^9$ (6) $18g^{-1}b^{-1}$ (7) $14r^{11}k^7$ (8) $4d^{-1}g^{-5}$

Exercise: Perform Division

9) $\dfrac{18g^3b^4}{9g^{-2}b^3}$

10) $\dfrac{6x^8y^8}{3x^3y^3}$

11) $\dfrac{7x^7y^8z^5}{7x^2y^5z^4}$

Answers: (9) $2g^5b$ (10) $6x^5y^5$ (11) x^5y^3z

Exercise: Solve the variable expressions (16 problems)

mathematics

Given K = 21. Then

1. 7 - K =
2. K - 3 =
3. 6 ÷ K =
4. K ÷ 2 =
5. K - 2 =
6. K - 1 = 7.
7 × K =
8. K ÷ 8 =
9. 3 + K =
10. K + 4 =

Given L = 4 then

9. L ÷ 5 =
10. L - 7 =
11. 2 - L =
12. 7 - L =
13. 9 - L =
14. L × 6 =
15. 6 - L =
16. L - 5 =
19. L + 4 =
20. L ÷ 9 =

Solving equations

Using the basic rule of equations we can find the value of variable when we know value of equation. This process is not entirely new. We have already done few above. You have read in previous pages and know how we transformed these equations

1) If $A + 8 = 14$ then $A = 14 - 8 = 6$
2) If $A - 10 = 20$ then $A = 20 + 10$
3) If $A \times 7 = 140$ then $A = \frac{140}{7} = 20$
4) If $\frac{A}{12} = 4$ then $A = 4 \times 12 = 48$

Exercise: Find the variable present in the given equation (100 problems)

1. $x \times 6 = 36$

2. a × 15 = 30

3. 2 × z = 12

4. 8 × x = 32

5. 4 × a = 12

6. a × 3 = 27

7. z × 5 = 45

8. x × 1 = 1

9. 6 × a = 30

10. 7 × a = 49

Exercise: Find the variable present in the given equation (73 problems)

1. 54 = 7a + 5

2. 28 = 6x + 4

3. 7 + 5h = 22

4. 33 = 9 + 3k

5. 56 = 7 + 7x

6. 24 = 6 + 6a

7. 8x + 5 = 13

8. 6 + 4k = 22

9. 4x + 2 = 10

10. 2 + 1h = 10

Algebraic terms

Algebraic expressions are formed with algebraic terms. A term may have any number of variables.

8 has zero number of variable

mathematics

'y' has only one variable

xy² This term has two variables x and y, because this is product of two separate terms x and y².

The sum of all exponents present in the term is called degree of the term. For example in the term $x^3 y^4 z^5$ the sum of exponents is 12. So, 12 is the degree of term.

Term	Degree	Number of variables
a^3	3	1
$a\,t^{-2}$	1 - 2 = 1	2
$x^3 y^4 z^5$	3 + 4 + 5 = 12	3

$8z^3$	0 + 3 = 3	1 and one constant
34	0	Constant

Like and unlike terms in polynomials

If the variable part of few terms are same, they are like terms. You can add or subtract them.

Example: Add 2x and 3y

you can add them and get single term, because they both are unlike terms.

So, 2x + 3y = (2x + 3y)

Example: Add 2ab and 3ab

The variable is ab in both terms. So they are like terms.

2ab + 3 ab = 5 ab.

Example: $4x^5 + 3x^3 + 4xy + xy^5 + 5xy + x^5 + 7x^3 + 10xy^5$

Solution:

First group all like terms as under.

$(4x^5 + x^5) + (3x^3 + 7x^3) + (4xy + + 5xy) + (xy^5 + 10xy^5)$

Now add

$= 5x^5 + 10x^3 + 9xy + 11xy^5$ .... Ans

Example: Simplify $-4b - [3a + \{3b - 2a + (2b - 7)\} - 4a + 5]$

$-4b - [3a + \{3b - 2a + 2b - 7\} - 4a + 5]$

$= -4b - [3a + 3b - 2a + 2b - 7 - 4a + 5]$

$= -4b - [3a - 2a - 4a + 3b + 2b - 7 + 5]$

$= -4b - [-3a + 5b - 2]$

$= -4b + 3a - 5b + 2$

$= 3a - 4b - 5b + 2$

$= 3a - 9b + 2$

Monomials

It is an algebraic expression consisting of one term.

If the term is comprised of positive powers then we can call it monomial. Thus a^3, $x^3 y^4 z^5$, $8z^3$ are monomials. But a t^{-2}, is not a monomial because the term has negative powers.

A monomial can have any number of variables.

A monomial's all exponents must be whole numbers. That means that 8, 3x, yx^2, $12p^3 q4 r$, 7x, 23 x^{15}, 2lm all are examples of monomials whereas 2lm - 3, $\frac{8z}{x}$, $12x^{-2}$, $2pq-24+ y$, a^x, $\sqrt{x}$ are not monomials, since these numbers don't fulfill all criteria.

mathematics

The degree of monomial is the highest power that one of its variable has.

2lm – 3 is not monomial because it has more than one term.

$\frac{8z}{x}$ is not monomial because it has a variable 'x' with negative power.

$12x^{-2}$ is not monomial because it has a variable 'x' with negative power

(2pq−24 + y) is not monomial because it has a more than one term.

√x is not monomial because it has the power of x is not whole number.

a^x is not monomial because it has a variable 'x' as power.

A monomial has one term

A binomial has two terms

A trinomial has 3 three terms

A quadratic has 4 terms

A Binomial

A binomial contains two monomials

A trinomial

A trinomial contains three monomials

A Polynomial

If it contains more than three monomials, then is it polynomial. Polynomial is a general term. In fact Monomials, binomials trinomials can also be generally called as polynomials.

Identify the polynomial-type for each.

1) $-xq^5 - 5b^3 + 6k^2c^4 + 9n^6z^4 + 3x^2k^7 - 7g$

   Ans: A polynomial with 6 Terms

2) $-5b^3 + 6k^2c^4 + 9n^6z^4 + 3x^2k^7$

   Ans: A polynomial with 4 Terms

3) $zq^3 + 8x^6 - 4r^4h^7$ **Ans:** Trinomial

4) $3n^5c^5b^4 + 6r^3 + 2z^2q^6$ **Ans:** Trinomial

5) $4y^7 + s$ **Ans:** Binomial

6) $kp^2 - 7s^7 + 9z^3g^6 - 8c^4y^6 + 4k^3z^5$

   Ans: A polynomial with 5 Terms

7) $-9k$ **Ans:** Monomial

8) $-4y - 2r$ **Ans:** Binomial

9) $-7zx^6 - 2z^3$ **Ans:** Binomial

10) $-4y^7x^3$ **Ans:** Monomial

   11) $zp^4 + 8n^3$ **Ans:** Binomial

Degree of polynomial

The degree of a polynomial is the highest degree of its monomials with non-zero coefficients.

Identifying the degree for each polynomial.

1) $-7x^2y^3c^5 + 4z^7 + 7p^2g^6$ degree is 10

2) $5 + 78 + 98$ degree is 0

   3) $-4y^7k^2 + cg^3 - 6x^4z^6 + 8b^5p^7$ degree is 12

   4) $9ch^6q^2 - 3z^4 + 4z^3n^7 + 5n^3p^5 - 8b^4 + 2b$ degree is 10

   5) $6b^5z - 2s$ degree is 6

   6) $5rp^2 - 7y^6 + 4s^4z^7 - 8b^2z^7 + 2y^5s^4$ degree is 11

   7) $i - k + j - g - h - x$ degree is 1

mathematics

8) $-3nq - 6c^2$ degree is 2

9) $8sp^5 z^2 - 3k^4$ degree is 8

10) $xp + 7y - 3q - 7h$ degree is 2

Single - Variable polynomials.

Solve each equation.(20 problems)

1) $26 = 8 + j$

2) $3 + t = 8$

3) $15 + r = 23$

Answers (1) $j = 18$ (2) $t = 5$ (3) $r = 8$

Naughty question

A figure resembling a spiral is shown with 35 matches.

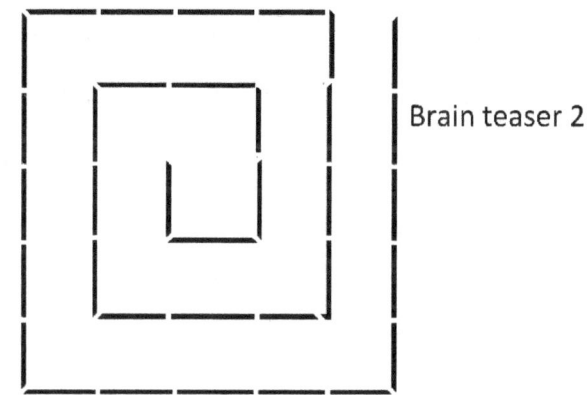

Brain teaser 2

Move 4 matches to form 3 squares.

4) $-15 + x = -9$

5) $x + 4 = -12$

6) $x - 7 = 13$

7) $m - 3 = -13$

8) $e - 5 = -5$

9) $w - 15 = -27$

10) $d + 16 = 9$

11) $-104 = 8f$

51

12) $14s = -56$

13) $-6 = 18d$

14) $10c = 40$

15) $3x - 12 = -27$

16) $4d + 15 = 7$

17) $104 = 52f$

18) $3 + 2t = 17$

19) $15 + 4r = 23$

20) $-15 + 6x = -9$

Answers: (4) x=6 (5) x=-16 (6) x=20 (7) m = -9 (8) e = 0 (9) w = -12 (10) d = -7 (11) f = -13 (12) s = -4 (13) d = -1/3 (14) c = 4 (15) x = -5 (16) d = -2 (17) f = 2 (18) t = 7 (19) r = 8 (20) x = 1

Distributive property of addition.

Addition is one of the four fundamental operations in mathematics. Students are familiar with two of its property.

First one is Commutative property. It says if there are two monomials to be added you can do it any way. For example

a) $2 + 3 = 5$ and $3 + 2 = 5$

b) $a + b = b + a$

Second one is associative property. It says if more than two monomials are to added you can select any number of variable in groups and add. For example

c) $8 + (7 + 5) = 20$ and $(8 + 7) + 5 = 20$

d) $(a + b) + c = a + (b + c)$

The third one is important.

If there is multiplication involved in addition the rule is you should do as follows.

$7(4 + 5) = 7(4) + 7(5) = 28 + 35 = 63$

$X(a + b) = Xa + Xb$

mathematics

This is called distributive property.

Adding polynomials of single variable and constants

Example:

$2x + 3(x + 10) + 9 - x$

$= 2x + 3x + 30 + 9 - x$

$= (2x + 3x - x) + (30 + 9) = 4x + 39$ ...... **Ans**

Exercise simplify the following expressions (5 problems)

$2x + 3x + 7x - 3x + 6$

$3(x + 4) + 2(x + 1)$

$4a + 8a + 3 + 7(a + 3)$

$2a + 23a + 34 + a =$

$a(3 + 5) + 4a + 10a =$

Exercise Solve for the variable (75 problems).

1. $43 - 5A = 8$
2. $11 = 3 - 8A$
3. $7A - 14 = 28$
4. $3A - 7 = 56$
5. $8A - 7 = 25$
6. $4A - 9 = 23$
7. $3A - 2 = 19$
8. $2 = 9 - 1A$
9. $3 = 27 - 8A$
11. $9 = 5A - 6$
12. $4 = 6A - 8$
13. $4 = 29 - 5A$
14. $22 - 2A = 4$
15. $9 = 16 - 7A$

16. 41 − 7A = 6
17. 2 = 3A − 7
18. 62 − 7A = 6
19. 13 − 1A = 5
20. 48 − 8B = 8
21. 3B − 7 = 2
22. 17 = 3B − 4
23. 2B − 4 = 2
24. 3B − 8 = 1
25. 6B − 5 = 37
26. 4 = 76 − 9B
27. 5 = 25 − 4B
28. 35 − 3B = 8
29. 13 − 6B = 1
30. 9B − 8 = 28
31. 29 = 6B − 7
32. 8B − 5 = 43
33. 6 = 24 − 9B
34. 9 − 1C = 0
35. 21 = 8C − 3
36. 8C − 4 = 60
37. 18 = 4C − 2
38. 17 = 6C − 1
39. 1 = 9 − 2C
40. 4 = 40 − 9C
41. 9C − 4 = 5
42. 5 = 77 − 8C
43. 16 = 8C − 8

mathematics

44. $2 = 14 - 2C$

45. $23 = 6C - 1$

46. $29 - 7C = 8$

47. $41 - 5C = 1$

48. $0 = 28 - 4C$

49. $5C - 6 = 29$

50. $38 = 8C - 2$

51. $7 = 23 - 4C$

52. $1C - 4 = 1$

53. $24 = 7C - 4$

54. $8 = 53 - 5C$

55. $8 = 7C - 6$

56. $35 - 8C = 3$

57. $9 = 17 - C$

58. $5 = 33 - 4C$

59. $20 = 3C - 1$

60. $6 = 26 - 4C$

61. $1 = 2C - 5$

62. $4 = 12 - 4X$

63. $3 = 66 - 7X$

64. $3 = 2X - 1$

65. $7 = 88 - 9X$

66. $3 = 39 - 6X$

67. $2 = 56 - 9X$

68. $3 = X - 3$

69. $5X - 1 = 29$

70. $3X - 6 = 3$

71. $3X + 1 = 16$

Algebra and Probability for kids

72. $27 - 3X = 0$

73. $5X - 7 = 3$

74. $4X + 4 = 16$

75. $42 - 3X = 0$

Exercise: Simplify each expression.(28 problems)

1) $(7z^2 - 8z) - (7z - 2z^2)$

2) $(5 - 9x^3 + 6x^4) + (3x^4 - 8x^2 + 4) + (2x^3 - 7x^2)$

3) $(4 - 8x^2 + 3x^4) + (7x + 5x^4 + 2)$

4) $(6(ab)^5 + 5) + (8(ab)^5 + 2 - 7(ab)^4) - (3(ab)^4 - 9(ab))$

5) $(g^5 + 7g^3 - 2) - (9g^5 - 4 + 8g^3) - (3g^3 + 6g + 5)$

6) $(2z^4 - 8z^3 - 9) - (5z^3 + 3z^2 - 6)$

7) $(9 - 8(ab)^2) + (6(ab)^3 + 2 - 7(ab)^2)$

8) $(6x^2 + 2x - 7) - (8x - 9) + (5x^4 + 3x^2 + 4x)$

9) $(7 + 6y^4 + 2y^2) + (5y^2 + 8y^3 - 9) - (y^4 - 3y^3 + 4)$

10) $4x^3 + 3x^4 - x^4 - 5x^3$

11) $5a + 4 - 5a + 3$

12) $3x^4 - 3x - 3x - 3x^4$

13) $3d^2 + 5d^2 - 3 + 2p^2 - 3p^3$

14) $6a^2 - 14a^3 + a^3 - 2a^2 - 3a^2 - 4a^3$

15) $4 + 2x^3 + 5x^3 + 2$

16) $4x - 3x^3 - 3x^3 + 4x$

17) $3a^2 + 1 - 4 + 2a^2$

18) $(-5y^3z^4 + 9y) + (-5y^3z^4 - 8y + 8y^2z^2) + (-8y^4z^2 + 8y^3z^4)$

19) $(-9xy^3 - 9x^4y^3) + (3xy^3 + 7y^4 - 8x^4y^4) + (3x^4y^3 + 2xy^3)$

20) $(y^3 - 7x^4y^4) + (-10x^4y^3 + 6y^3 + 4x^4y^4) - (x^4y^3$

mathematics

$+ 6x^4y^4)$

21) $-4(ab)^4 + 14 + 3(ab)^2 + -3(ab)^4 - 14(ab)^2 - 8$

22) $5 - 6x^5 - 8x^4 - -6x^4 - 5x - 8x^5$

23) $12a^5 - 6a - 10a^3 - 10a - 2a^5 - 14a^4$

24) $8x - 5x^4 + 10x^2 - 5x^2 + 11x^4 - 7$

25) $-x^4 + 15x^5 + 6x^3 + 6x^3 + 5x^5 + 7x^4$

26) $9x^3 + 5x^2 + 11x + -2x^3 + 9x - 8x^2$

27) $15x^2 + 11x - 2x^4 + -15x^2 - 3x - 6x^4$

28) $-7x^5 + 14 - 2x + 10x^4 + 7x + 5x^5$

Exercise: Simplify each expression(18 problems)

1) $5K^2 - 5 + 2K^2 - 5K^3$

2) $V^3 - 2V^2 - 3V^2 - 4V^3$

3) $4 + 2B^3 + 5B^3 + 2$

4) $4B - 5B^3 - 5B^3 + 4B$

5) $3V^2 + 1 - 4 + 2V^2$

6) $4r^3 + 3r^4 - r^4 - 5r^3$

7) $(5W + 4) - (5W + 3)$

8) $(7x^4 - 6x) - (6x - 7x^4)$

9) $(-3k^4 + 14 + 3k^2) + (-3k^4 - 14k^2 - 8)$

10) $3 - 6B^5 - 8B^4 - (-6B^4 - 3B - 8B^5)$

11) $(12W^5 - 6W - 10W^3) - (10W - 2W^5 - 14W^4)$

12) $(8B - 3B^4 + 10B^2) - (3B^2 + 11B^4 - 7)$

13) $(-x^4 + 13x^5 + 6x^3) + (6x^3 + 5x^5 + 7x^4)$

14) $(9r^3 + 5r^2 + 11r) + (-2r^3 + 9r - 8r^2)$

15) $(13B^2 + 11B - 2B^4) + (-13B^2 - 3B - 6B^4)$

16) $(-7x^5 + 14 - 2x) + (10x^4 + 7x + 5x^5)$

17) $(7 - 13x^3 - 11x) - (2x^3 + 8 - 4x^5)$

18) $(13a^2 - 6a^5 - 2a) - (-10a^2 - 6a^5)$

Word problems

Problem 1: Vihaan said to Ilo. Imagine a number and think so many dollars are there in your pocket.

I have none. Said Ilo. Vihaan said " I know that your dad will never give you a dime. But you can just assume a number, any number … just imagine.

Then Ilo thought for a moment. OK. I thought of some number. Shall I tell you how much?.... "NOOOOOO" shouted Vihaan. It is your secret. Do tell me. Now you have some dollars in your pocket. Go to money lender and take a loan of as many dollars as you have now in your pocket.

Ilo thought for a moment and said I took the loan. What now?

" You have double money now. You are rich. In addition I would like to give you 50 dollars more. Take them add to your property.

" That is very nice of you" Ilo said with joy and started adding 50 dollars to his total. He closed his eyes calculated and said" I added, what now?" Then Vihaan said you are very rich now by grace of God. You owe God a lot. Give half of money to God and calculate in your mind what is left thereafter.

"True" said Ilo…I gave half of what I had to GOD. I threw the money in James river and came back. What now?

Then Vihaan said. Give the money you borrowed before, back to money lender. "Oh!" said Ilo "I gave back the money I took from money lender. I am not that much rich now. Shall I tell you how much I have now.

No" said Vihaan. You need not. I know you have 25 dollars left with you"

Ilo wondered. How could you guess so accurately?

Can any one explain to Ilo how Vihaan could accurately tell the answer?

Solution:

Let us say Ilo assumed x dollars.

Then he took loan of another x dollars. $X + x = 2x$

Vihaan gave him 50 dollars. So Ilo had $(2x + 50)$ dollars now.

He then gave half of it to God. Then he has only half

$= ½ [2x + 50]$

Used distributive property of addition here and calculate

$½ [2x + 50] = ½ (2x) + ½ (50)$

$= x + 25$

He then gave back x dollars to money lenders

The total now is

$X + 25 - x = 25$

The final equation has no variable. The variable x has been eliminated. In such cases if Ilo assumes some other figure it gets eliminated finally and he will have only half of Vihaan gives him.

Problem 2:

A man wanted to go home and give flowers to his wife. He picked some flowers. Before reaching home he had to take bath in three water pools.

He got into first pool took his bath. When he came out he noticed that the flowers in his bag had doubled. He threw three flowers into pool as mark of gratitude.

He went to second pool and took bath. When he came out he noticed that the number of flowers in the bag doubled again. "Thanks! He told to pool and dropped five flowers in the pool and carried on to the third and final pool. After taking bath in the third pool had again noticed that the flowers have doubled. He gave ten flowers to the pool and stepped into hid house but there were no flowers in the bag to give to his wife.

Can any one tell how many flowers he had first, in his bag.

Solution:

Suppose the man started with x number of flowers in his bag.

First Pool

When he took bath in the first pool, they doubled.

$x \times 2 = 2x$ flowers. Then he left 3 flowers in the pool.

Remaining with him now are $(2x - 3)$ flowers.

Second pool

When he dipped in second pool the flowers doubled.

That means now he has $2 \times (2x - 3) = 2 \times 2x - 2 \times 3$

$= 4x - 6$ flowers.

Then he left 5 flowers in the pool.

mathematics

Remaining with him now are (4x - 11) flowers

Third pool

When he dipped in third pool the flowers doubled. That means now has $2 \times (4x - 11) = 2 \times 4x - 2 \times 11 = 8x - 22$ flowers.

Then he left 10 flowers in the pool.

Remaining with him now are (8x - 32) flowers

Problem 3: Few birds were flying in the sky. And few birds are sitting on ground. A bird sitting on floor said to birds flying in the sky "If one of you come down and join us we both have same number of birds." Then the birds in the sky said to the birds sitting on ground " If one of you fly and join us we will be double in number to you sitting on ground" How many birds are flying and how many birds are sitting on the ground?

Solution:

Let the birds sitting on the ground be x and the number of birds flying in the sky be y.

If one flying bird comes down and join sitting birds, the numbers will be

(x + 1) on the ground and (y - 1) on the sky.

But it is said x + 1 = y - 1 or x = y - 2 ----- (1)

If one bird from the flock sitting down flies and join the flock of flying birds the number will be

(x - 1) on the ground and (y + 1) in the sky.

But it says

Y + 1 = 2(x - 1) or y + 1 = 2x - 2

Y = 2x - 3

Now substitute x = (y − 2) from (1)

Y = 2(y -2) - 3 = 2y − 4 - 3 = 2y - 7

Or y − 7 = 0 which means y = 7

Substitute this in (1)

X = 7 − 2 = 5

So there are 5 birds sitting on the ground and 7 birds flying....

Ans

Problem 4: The sum of the digits in a two-digit number is 5. If we interchange the digits then the new number formed is 9 less. Find the original number.

Solution:

A two digit number should be represented in algebra as 10x + y

that means x is in place value tens and y is in place value units.

If we interchange the digits we get a new number that is represented as 10 y + x

Sum of numbers is given as 5

⇔ x + y = 5

The new number is 9 less than original number.

⇔ 10y + x − (10x + y) = 9

Problem 5: There are 39 boys in the class. This is three more than four times the number of girls. How many girls are in the class?

Solution:

Let x be the number of girls.

Given $39 = 4x + 3$

Subtracting 3 from both sides we have

$39 - 3 = 4x + 3 - 3$

$\iff 36 = 4x$

Now divide both sides by 4

$\iff \dfrac{36}{4} = \dfrac{4x}{4}$

$\iff x = 9$

There are 9 girls in the class.

Problem 5a: A rooster has laid eggs 10 times to the number of its legs. How many eggs did it lay?

Solution:

Kidding? Can a rooster lay eggs? What is that algebra has to do with it?

Problem 6: The sum of two numbers is 40, and one of them is 14 more than the other. What are the two numbers?
Solution:

Let one number is x

Then other number is x + 14

Given x + x + 14 = 40

⇔ 2x + 14 = 40

⇔ 2x = 40 − 14 = 26

⇔ x = 13

⇔ the other number is 13 + 14 = 27 .... **Ans**

Problem 7: Half a number plus 5 is 43. What is the number?

Solution

Let x be the number.

given $\frac{x}{2} + 5 = 43$

⇔ $\frac{x}{2} + 5 - 5 = 43 - 5$

⇔ $\frac{x}{2} = 38$

⇔ $2 \times \frac{x}{2} = 38 \times 2$

⇔ x = 76 ... **Ans**

Problem 8: The sum of two consecutive integers is 35. What are the two numbers?

Solution

Let n be the first integer and

mathematics

let n + 1 be the second integer

n + n + 1 = 35

2n + 1 = 35

2n + 1 - 1 = 35 - 1

2n = 34

n = 17: This is the first integer

Second one is 18......... **Ans**

Problem 9: The sum of two consecutive even integers is 30. What are the two numbers?

Solution

Let 2n be the first even integer and let 2n + 2 be the second integer

2n + 2n + 2 = 30

4n + 2 = 30

4n + 2 - 2 = 30 - 2

4n = 28

2n = 14: This is the first even integer

Second one is 2n + 2 = 14 + 2 = 16......... **Ans**

Problem 10: The sum of two numbers is 21. The difference is 3. What are the two numbers?

Solution:

Let x, y be the two numbers

Given that $x + y = 21$

$\Leftrightarrow y = 21 - x$

Given that $x - y = 3$

Eliminate y in above equation

$x - y = 3 \Leftrightarrow x - (21 - x) = 3$

$\Leftrightarrow x - 21 + x = 3$

$\Leftrightarrow -21 + 2x = 3$

Adding 21 the left side and the right side gives,

$2x = 24$

$x = 12$

Since $x + y = 21$, $10 + y = 16$

$12 + y = 21$

$12 + y - 12 = 21 - 12$

$y = 9$

The numbers are 12 and 9 ……….. **Ans**

Problem 11: six years ago, Victor's age was half of the age he will be in 10 years. How old is he now?

Solution:

Let x be Victor's age now.

6 years ago Victor's age was $x - 6$

His age after 10 years will be $x + 10$

Given $2 \times (x - 6) = x + 10$

Let us transform this equation to find x

Apply distributive property. That makes

$2x - 12 = x + 10$

mathematics

2x − 12 − 10 = x + 10 − 10

2x − 22 = x

2x − 22 − x = x − x

x − 22 = 0

x − 22 + 22 = 22

x = 22

Victor is now 22 years old .... **Ans**

Problem 12: A bat and a ball cost one dollar and ten cents in total. The bat costs a dollar more than the ball. How much does the ball cost?

Solution:

Let us do this problem using unit 'cent'.

Let cost of ball be x cents.

Then cost of bat would be x + 100 cents.

Given that x + x + 100 = 110 cents

2x + 100 = 110 cents

2x + 100 − 100 = 110 − 100

2x = 10 cents

X = 5 cents

So the cost of ball is 5 cents and cost of bat is 105 cents . . . **Ans**

Problem : It took 20 days for 20 men to build a wall. In how many days can 100 people build the same wall?

Solution:

Kidding? It has nothing to do with algebra. The wall is already built.

Algebra and Probability for kids

Problem 13: The sum of two numbers is 51. One of the numbers exceeds the other by 9. Find the numbers.

Solution:

Let the number be x.

Then the other number = (x + 9)

Sum of two numbers = 51

According to question, x + x + 9 = 51

⇔ 2x + 9 = 51

⇔ 2x = 51 - 9

⇔ 2x = 42

⇔ 2x/2 = 42/2 (divide by 2 on both the sides)

⇔ x = 21

Therefore, x + 9 = 21 + 9 = 30

Therefore, the two numbers are 21 and 30.

Problem 14: The difference between the two numbers is 54. The ratio of the two numbers is 5: 4. What are the two numbers?

Solution:

Let the common ratio be x. Then we can assume that the two numbers are 5x and 4x.

Their difference given as 54

5x - 4x = 54

mathematics

$\Leftrightarrow x = 54$

$\Leftrightarrow x = 48/4$

$\Leftrightarrow x = 12$

Therefore, $5x = 5 \times 54 = 270$

$4x = 4 \times 54 = 216$

Therefore, the two numbers are 216 and 270.

Problem 15: The length of a rectangle is twice its breadth. If the perimeter is 96 meter, find the length and breadth of the rectangle.

Solution:

Let the breadth of the rectangle be x,

Then the length of the rectangle = 2x

Perimeter of the rectangle = 96

Therefore, according to the question
$2(x + 2x) = 96$

$\Leftrightarrow 2 \times 3x = 96$

$\Leftrightarrow 6x = 96$

$\Leftrightarrow x = 96/6$

$\Leftrightarrow x = 16$

We know, length of the rectangle = 2x

$\qquad = 2 \times 16 = 32$

Therefore, length of the rectangle is 32 m and breadth of the rectangle is 16 m.

Problem 16: Vicky is 5 years younger than Ricky. Four years later, Ricky will be twice as old as Vicky. Find their present ages.

Solution:

Let Ricky's present age be x.

Then Vicky's present age = x - 5

After 4 years Ricky's age = x + 4, Vicky's age x - 5 + 4.

Given that Ricky will be twice as old as Vicky.

Therefore, x + 4 = 2(x - 5 + 4)

$\Leftrightarrow$ x + 4 = 2(x - 1)

$\Leftrightarrow$ x + 4 = 2x - 2

$\Leftrightarrow$ x + 4 = 2x - 2

$\Leftrightarrow$ x - 2x = -2 - 4

$\Leftrightarrow$ -x = -6

$\Leftrightarrow$ x = 6

Therefore, Vicky's present age = x - 5 = 6 - 5 = 1

mathematics

Therefore, present age of Ricky = 6 years and present age of Vicky = 1 year.

Binomials: The polynomial with two monomials is called by binomial.

x + y is a binomial.

Adding binomials

Example:

Add 2x + 3y and x – y

Solution:

In these binomials all x terms are like terms and all y terms are like terms. So they can be added. The addition can be done both in vertical way and also horizontal way.

Horizontal way

(2x + 3y) + (x - y)

Open the brackets

2x + 3y + x – y

Group like terms

2x + x + 3y – y

= 3x + 2y .... **Ans**

Vertical way

Put two expressions one under the other and add the like terms.

$$2x + 3y$$
$$X - y$$
$$= 3x + 2y ...$$

Ans

Example: Add $2x^2y + 3a^2b$ and $x^2y - a^2b$

Solution:

In these binomials all (x^2y) terms are like terms and all (a^2b) terms are like terms. So they can be added. The addition can be done both in vertical way and also horizontal way.

Horizontal way

$(2x^2y + 3a^2b) + (x^2y - a^2b)$

Open the brackets

$2x^2y + 3a^2b + x^2y - a^2b$

Group like terms

$2x^2y + x^2y + 3a^2b - a^2b$

$= 3x^2y + 2a^2b$ .... **Ans**

Vertical way

Put two expressions one under the other and add the like terms.

$$2x^2y + 3a^2b$$
$$x^2y - a^2b$$
$$= 3x^2y + 2a^2b \ldots$$

Ans

Example: Subtract $2x - 4y$ from $5x + 8y$

Solution:

In these binomials all x terms are like terms and all y terms are like terms. So they can be added. The addition can be done both in vertical way and also horizontal way.

Horizontal way

(5x + 8y) - (2x − 4y)

Open the brackets

5x + 8y - 2x + 4y

Group like terms

5x - 2x + 8y + 4y

= 3x + 12y .... **Ans**

Vertical way

Put two expressions one under the other and add the like terms.

$$5x + 8y$$
$$\underline{-(2x - 4y)} = -2x + 4y$$

Now add

= 3x + 12y ... **Ans**

Example: Subtract $x^2y^3 - a^2b$ from $2x^2y^3 + 3a^2b$

Solution:

In these binomials all ($x^2 y^3$) terms are like terms and all (a^2b) terms are like terms. So they can be added or subtracted. The Subtraction can be done both in vertical way and also horizontal way.

Horizontal way

($2x^2 y^3 + 3 a^2b$) - ($x^2 y^3 - a^2b$)

Open the brackets. Notice the second expression which changes its sign.

$2x^2y^3 + 3a^2b - x^2y^3 + a^2b$

Group like terms

$2x^2y^3 - x^2y^3 + 3a^2b + a^2b$

$= x^2y^3 + 4a^2b$ .... **Ans**

Vertical way

Put two expressions one under the other and add the like terms.

$$2\ x^2y^3 + 3a^2b \ldots (1)$$
$$-\ (x^2y - a^2b) \text{---} (2)$$
$$= -x^2y + a^2b \text{ (change the sign)}$$
$$\text{Add}\ \ x^2y + 4a^2b \ldots \textbf{Ans}$$

Exercise (40 problems) Simplify each expression.

1) $(x + y) + (x - y)$
2) $(3x + 4y) + (5x - 6y)$
3) $(5x + 6y) - (8x - 5y)$
4) $(3x + 6y) + (7x - 8y)$
5) $(6x + 9y) - (5x - 7y)$
6) $(2x + 6y) + (5x - 9y)$
7) $(9x + 4y) + (5x - 6y)$
8) $(2x + 6y) - (x - 5y)$
9) $(8x + 7y) + (2x - 8y)$
10) $(4x + 9y) - (5x - 2y)$
11) $(5w^2 - 3) + (2w^2 - 3w^3)$
12) $(x^3 - 2x^2) - (3x^2 - 6x^3)$

mathematics

13) $(6 + 2n^3) + (5n^3 + 2)$

14) $(6n - 3n^3) - (3n^3 + 6n)$

15) $(3x^2 + 1) - (6 + 2x^2)$

16) $(6r^3 + 3r^6) - (r^6 - 5r^3)$

17) $(5x + 6) - (5x + 3)$

18) $(3x^6 - 3x) - (3x - 3x^6)$

19) $[-6(uv)^6 + 16 + 3(uv)^2] + [-3(uv)^6 - 16(uv)^2 - 8]$

20) $(3 - 6n^5 - 8n^6) - (-6n^6 - 3n - 8n^5)$

21) $(12x^5 - 6x - 10x^3) - (10x - 2x^5 - 16x^6)$

22) $(8n - 3n^6 + 10n^2) - (3n^2 + 11n^6 - 7)$

23) $(-x^6 + 13x^5 + 6x^3) + (6x^3 + 5x^5 + 7x^6)$

24) $(9r^3 + 5r^2 + 11r) + (-2r^3 + 9r - 8r^2)$

25) $(13n^2 + 11n - 2n^6) + (-13n^2 - 3n - 6n^6)$

26) $(-7x^5 + 16 - 2x) + (10x^6 + 7x + 5x^5)$

27) $(7 - 13x^3 - 11x) - (2x^3 + 8 - 6x^5)$

28) $(13x^2 - 6x^5 - 2x) - (-10x^2 - 11x^5 + 9x)$

29) $(3x5 + 8x^3 - 10x^2) - (-12x^5 + 6x^3 + 16x^2)$

30) $(8b^3 - 6 + 3b^6) - (b^6 - 7b^3 - 3)$

31) $[(uv)^6 - 3 - 3(uv)^3] + [-5(uv)^6 + 6(uv)3 - 8(uv)^5]$

32) $[-10(ab)^2 + 7(ab) + 6(ab)^6] + [-16 - 6(ab)^6 - 16(ab)]$

33) $(-7n^2 + 8n - 6) - (-11n + 2 - 16n^2)$

34) $(16w^6 + 11w^2 - 9w^5) - (-2w^6 + 5w^5 - 11w^2)$

35) $(8(ab) + (ab)^2 - 6) - (-10(ab) + 7 - 2(ab)^2)$

36) $(-9x^2 - 8y) + (-2yx - 2y^2 + x^2) + (-x^2 + 6yx)$

37) $(4x^2 + 7x^3y^2) - (-6x^2 - 7x^3y^2 - 4x) - (10x + 9x^2)$

38) $(-5y^3x^4 + 9y) + (-5y^3x^4 - 8y + 8y^2x^2) + (-8y^4x^2 + 8y^3x^4)$

39) $(-9xy^3 - 9x^4y^3) + (3xy^3 + 7y^4 - 8x^4y^4) + (3x^4y^3 + 2xy^3)$

40) $(y^3 - 7x^4 y^4) + (-10x^4 y^3 + 6y^3 + 4x^4 y^4) - (x^4 y^3 + 6x^4 y^4)$

Solving linear equations with two variables

Example Solve the equations $2x + 3y = 7$ and $x + 2y = 8$

Solution:

$2x + 3y = 7$ .... (1)

$x + 2y = 8$ .... (2)

We must eliminate x or y from these equation

Let us eliminate y

To do that first transform the two equations so that coefficients of y are same in both equations.

Multiply equation 1 with 2

$\Leftrightarrow 2(2x + 3y) = 2 \times 7$

$\Leftrightarrow 4x + 6y = 14$

Multiply equation 2 with 3

$\Leftrightarrow 3(x + 2y) = 3 \times 8$

$\Leftrightarrow 3x + 6y = 24$

We have now two transformed equations as under

$4x + 6y = 14$ .... (1a)

$3x + 6y = 24$ .... (2a)

To eliminate y, subtract 2a from 1a

You get x = -10

Now we know the value of x. to know 'y' substitute value of x either in 1 or 2

I want to do with equation 1

mathematics

$2x + 3y = 7 \Leftrightarrow -20 + 3y = 7$

$3y = -27$

$y = -9$ .... Ans

Solve these Equations

$2x + 3y = 5$ and $3x + 4y = 7$

Solution: $x = 1$ and $y = 1$

$x + 3y = 11$ and $3x + y = 9$

Solution: $x = 2$ and $y = 3$

$5x - 4y = 1$ and $2x + 5y = 7$

Solution: $x = 1$ and $y = 1$

$3x - 5y = -1$ and $3x - 4y = 8$

Solution: $x = 3$ and $y = 2$

$2x + 3y = -5$ and $3x + 4y = -7$

Solution: $x = -1$ and $y = -1$

$10x + y = 35$ and $x + 10y = 53$

$X = 3, y = 5$

$5x - 7y = 3$ and $3x + 4y = 1$

Solution: $x = -5$ and $y = 4$

$2x + 3y = 0$ and $4x + 5y = 2$

Solution: $x = 3$ and $y = -2$

Multiplication of binomials

Multiplying a binomial with a monomial is done using property of distribution.

Distribution property of multiplication as already mentioned is

$A(B + C) = AB + AC$

Where A is monomial and $B + C$ is binomial.

Example: $7x^3 (5x + y^2)$

Solution:

Use distributive property of multiplication.

$7x^3 (5x^2 + y^2)$

$= 7x^3 \times 5x^2 + 7x^3 \times y^2$

$= 35x^5 + 7x^3 y^2$ .... **Ans**

Example: $-7a (4a - 6b)$

Solution:

Use distributive property of multiplication.

$-7a (4a - 6b)$

$= (-7a) \times (4a) - (-7a) \times (6b)$

$= -28a^2 - (-42 ab)$

$= -28a^2 + 42 ab$............. **Ans**

Multiplying a binomial with another binomial is done using property of distribution.

Distribution property of multiplication as already mentioned is

$A(B + C) = AB + AC$

Now $(A + B) \times (C + D)$ can be done as follows where $(A + B)$ and $(C + D)$ are both binomials..

Example: $(6y + 3)(6y - 4)$

Solution:

Let us assume $6y + 3 = K$

Then given expression becomes

mathematics

K(6y – 4)

= 6 y K – 4K

Now substitute 6y – 3 for K

6y (6y – 4)   - 4 (6y – 4)

= 36 y² – 24 - 24 y + 16

= 36 y² - 24 y – 8 .... **Ans**

Example: Expand (3p + 2q)²

Solution:

(3p + 2q)² = (3p + 2q) (3p + 2q)

Let us write one 3p + 2q as K

Then given expression becomes

K (3p + 2q)

= 3 p K + 2 q K

Now substitute 3p + 2q for K

3 p (3p + 2q) + 2 q (3p + 2q)

= 9 p² + 6 pq + 6 pq + 4 q²

= 9 p² + 12 pq + 4 q² .... **Ans**

Example: Expand (a + b)²

Solution:

(a + b)² = (a + b)(a + b)

= a (a + b) + b(a + b)

= a² + ab + ab + b²

= a² + 2ab + b²

Example: Expand (a - b)²

Solution:

(a - b)² = (a - b)(a - b)

$= a(a - b) - b(a - b)$

$= a^2 - ab - ab + b^2$

$= a^2 - 2ab + b^2$

Example: Expand (a + b)(a − b)

Solution:

$(a + b)(a - b)$

$= a(a - b) + b(a - b)$

$= a^2 - ab + ab - b^2$

$= a^2 - b^2$

Exercise: Find product. (28 problems)

1) $6x(2x + 3)$

2) $2(-5x - 8)$

3) $2z(-2z - 3)$

4) $-4(z + 1)$

5) $(2y + 2)(6y + 1)$

6) $(4y + 1)(2y + 6)$

7) $(z - 3)(6z - 2)$

8) $(8p - 2)(6p + 2)$

9) $(6p + 8)(5p - 8)$

10) $(3m - 1)(8m + 2)$

11) $(2a - 1)(8a - 5)$

12) $(5y + 6)(5y - 5)$

13) $(4p - 1)^2$

14) $(2z - 6)(5z + 6)$

15) $(6y + 3)(6y - 4)$

16) $(8y + 1)(6y - 3)$

mathematics

17) $(6k + 5)(5k + 5)$

18) $(3z - 4)(4z + 3)$

19) $(4a + 2)(6a^2 - a + 2)$

20) $(2k - 3)(k^2 - 2k + 2)$

21) $(2r^2 - 6r - 6)(2r - 4)$

22) $(y^2 + 6y - 4)(2y - 4)$

23) $(y - 5)(2y^2 + 6y - 5)$

24) $(2n - 5)(2n^2 - 3n - 2)$

25) $(a + b)(a^2 - ab + b^2)$

26) $(a - b)(a^2 + ab + b^2)$

27) $(a + b)^2 - (a - b)^2$

28) $(a + b)^2 - (a - b)^2$

Example 1: There are two whole numbers. The sum of twice one number and three times another number is 23 and their product is 20. Find the numbers.

Solution:

Let the numbers be x and y

The sum of twice a number and three times another number is 23.

$\Leftrightarrow 2x + 3y = 23$

Their product is 20.

$xy = 20$

Rearranging the first equation gives

$(3y) = 23 - 2x$

$Y = \dfrac{23 - 2x}{3}$

We know y now. substituting this the second gives

$$x \times \frac{23 - 2x}{3} = 20$$

Multiplying each side of the equation by 3 gives

$23x - 2x^2 = 60$

Rewriting this equation in standard quadratic form gives

$2x^2 - 23x + 60 = 0$

From here we use **distributive property in reverse**.

$2x^2 - 23x + 60 = 0$

$\Leftrightarrow 2x^2 - 15x - 8x + 60 = 0$

$\Leftrightarrow [2x^2 - 15x] - [8x - 60] = 0$

$\Leftrightarrow x \times [2x - 15] - 4[2x - 15] = 0$

Let $2x - 15 = K$ then equation becomes

$Kx - 4K = K(x - 4) = 0$

Or $(2x - 15)(x - 4) = 0$

If the product of two terms is zero, either one of them must be zero. Both the terms are made of single variable x. Hence they both can not be zero.

If we believe $2x - 15 = 0$ and solve for x,

we get $x = 7\frac{1}{2}$ which is not a whole number. So it is ignored

If $x - 4 = 0$, $x = 4$

That's is the answer. One of the two numbers

is 4 and other is 5 .... **Ans**

Example 2: A triangle has a perimeter of 65cm. If 2 of its sides are equal and the third side is 5cm more than the equal sides, what is the length of the third side?

Solution:

mathematics

Let length of the equal side $= x$

Given that $65 = x + x + (x + 5)$

$65 = 3x + 5$

$3x = 65 - 5$

$3x = 60$

$x = 20$

The length of third side $= 20 + 5 = 25$cm .... **Ans**

Example 3: A rectangle is 4 times as long as it is wide. If the length is increased by 4 inches and the width is decreased by 1 inch, the area will be 60 square inches. What were the dimensions of the original rectangle?

Solution:

Let $x =$ original width of rectangle

$l = 4x$ and $b = x$

Area formula $A = l\,b$

Given that $60 = (4x + 4)(x - 1)$

Use distributive property to remove brackets

$\Leftrightarrow 60 = 4x^2 - 4x + 4x - 4$

$\Leftrightarrow 4x^2 - 4 - 60 = 0$

$\Leftrightarrow 4x^2 - 64 = 0$

$\Leftrightarrow 4x^2 - 64 + 64 = 64$

$\Leftrightarrow 4x^2 = 64$

Divide both sides by 4

$\Leftrightarrow x^2 = 16$

Take square root both sides

$\Leftrightarrow x = +4$ or -4

Width can not be negative, So x = 4

Then length = 4 × 4 = 16 .... **Ans**

Example 4: Orion's father is 5 times older than Orion and Orion is twice as old as his sister Rema. In two years time, the sum of their ages will be 58. How old is Orion now?

Solution:

Let Rema's age = x

Orion's age = 2x

Father's age = 5 × 2x = 10 x

Sum of ages after two years

= (x + 2) + (2x + 2) + (10 x+2) = 58

$\Leftrightarrow$ 13x + 6 = 58

$\Leftrightarrow$ 13x = 52

x = 4

So Rema's age = 4, Orion's age = 8

and their father's age = 40 ---------------- **Ans**

Example 5: In the given parallelogram ABCD, find CD.

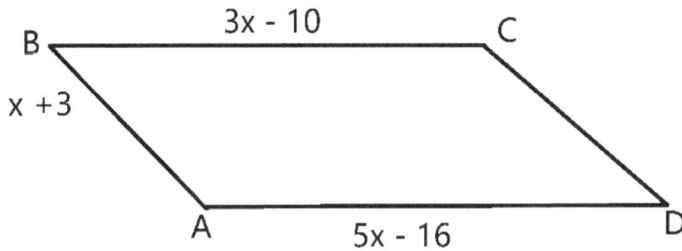

Solution:

mathematics

$3x - 10 = 5x - 16$ (opposite sides)

$\Leftrightarrow 3x - 5x = -16 + 10$

$\Leftrightarrow -2x = -6 \quad \Leftrightarrow x = 3$

Since $x = 3$, $AB = x + 3 = 6$

$\therefore CD = AB = 6$ .... **Ans**

Example 6: A swarm of 45 butterflies are moving in a garden. 5 butterflies sit on each flower. But there are no flowers left for 10 butterflies. Find the number of flowers.

Solution:

Let the number flower = x

Then $5x + 10 = 45$

$5x = 45 - 10 = 35$

$X = \dfrac{35}{5} = 7$ .... **Ans**

Exercise in solving linear equations with two variables

1) $y = \dfrac{x}{2} + 6$

$y = 2$

2) $y = \dfrac{x}{3} - 4$

$$y = \frac{7x}{4} + 4$$

3) $y = \frac{x}{2} + 3$

$y = 2x - 2$

4) $6x + y = 15$

$-7x - 2y = -10$

5) $2x + 3y = -12$

$-x - 3y = 18$

6) $y = \frac{3x}{4} - 3$

$y = -2x - 3$

7) $-3x + 2y = 23$

$5x + 2y = -17$

8) $y = 9x - 9$

$y = 9$

9) $y = 7x - 3$

$y = 4$

1. $3x + z = 15$ and $x + 2z = 10$

2. $x + 6y = 32$ and $x + 3y = 17$

3. $3c+4u = 33$ and $6c+3u = 36$

4. $6x+y = 18$ and $5x+2y = 22$

5. $2a+2x = 18$ and $a+3x = 17$

6. $5a+2b = 32$ and $6a+6b = 42$

7. $5a+b = 15$ and $2a+6b = 34$

8. $2x+3y = 8$ and $3x+6y = 15$

9. $x+5y = 35$ and $4x+2y = 32$

10. $2x+6y = 42$ and $2x+4y = 30$

11. $4x+3y = 29$ and $6x+3y = 39$

12. $2a+6b = 16$ and $6a+b = 31$

Answers

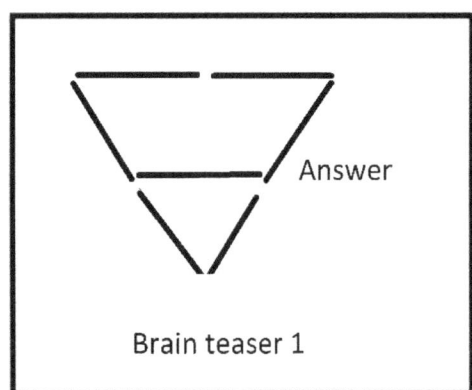

Brain teaser 1

A figure resembling a spiral is shown with 35 matches.

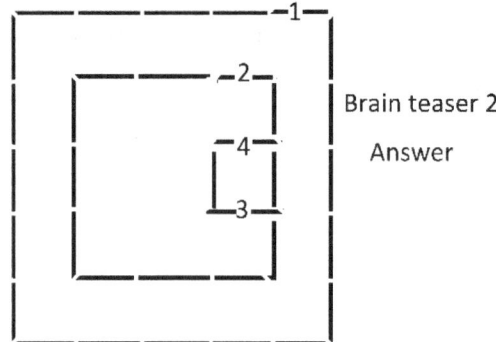

Brain teaser 2

Answer

Move 4 matches to form 3 squares.

mathematics

Probability for Kids

What is probability?

Probability is the measure of how likely a required event shows up when there is chance for more number of other event to appear.

Many events can't be predicted accurately. The best we can say is, how likely they will happen, using the idea of probability.

In mathematics, the extent to which an event is likely to occur is measured by the ratio of the favorable cases to the whole number of cases.

Suppose we have a die. It has 6 faces. Any face can come up. Total number of cases here are 6

Suppose we have a coin. Coin has two faces. Heads and tails. If we toss a coin any face can show up. Total number of cases here are two.

An experiment is a situation involving chance or probability that leads to results called outcomes. An outcome is the result of a single trial of an experiment.

An event is required number of outcomes of an experiment.

Probability Key Terms

Experiment

An experiment in probability is a test to see what will happen incase you do something. A simple example is flipping a coin. When you flip a coin, you are performing an experiment to see what side of the coin you'll end up with.

mathematics

Outcome

An outcome in probability refers to a single (one) result of an experiment. In the example of an experiment above, one outcome would be heads and the other would be tails.

Event

An event in probability is the set of a group of different outcomes of an experiment. Suppose you flip a coin multiple times, an example of an event would the getting a certain number of heads.

Sample Space

A sample space in probability is the total number of all the different possible outcomes of a given experiment. If you flipped a coin once, the sample space S would be given by:

S = 2, = { head, tail }

If you flipped the coin multiple times, all the different combinations of heads and tails would make up the sample space.

Notation of Probability

The probability that a certain event will happen when an experiment is performed can in layman's terms be described as the chance that something will happen.

The probability of an event, E is denoted by P(E)

Suppose that our experiment involves rolling a die. There are 6 possible outcomes in the sample space, as shown below:

Algebra and Probability for kids

S = 6 = {1, 2, 3, 4, 5, 6}

The size of the sample space is often denoted by N while the number of outcomes in an event is denoted by n.

From the above, we can denote the probability of an event as:

P of event =

$$\frac{\text{required number or favorable number of out comes}}{\text{Total number of out comes in sample space}}$$

Worked out examples.

Example1: What is probability of striking red color in the spinning wheel shown.

mathematics

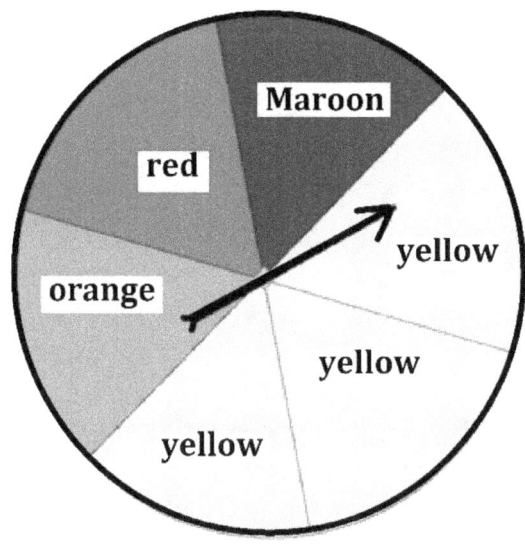

Solution:

P of event =

$$\frac{\text{required number}}{\text{Total number of out comes in sample space}}$$

The total number of outcomes that can occur = 6

The event which we are expecting to come is just 1

Hence P(red color) = 1/6 .... **Ans**

Example 2: What is probability of striking Yellow color in the spinning wheel shown.

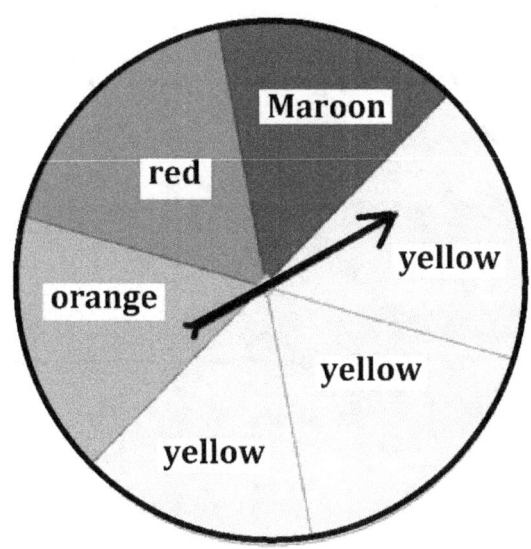

Solution:

P of event =

$$\frac{\text{required number}}{\text{Total number of out comes in sample space}}$$

The total number of outcomes that can occur = 6

The events which we require, is one of 3

So,

Hence P(Yellow color) = 3/6 = ½   .... **Ans**

Example 3: A Spinner has four equal sectors with orange, blue, red and green colors. What is probability of landing on blue color on the spinning wheel, when it is spun.

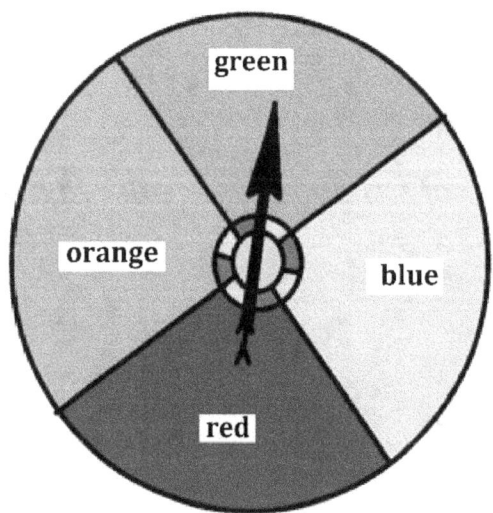

Solution:

P of event =

mathematics

$$\frac{\text{required number}}{\text{Total number of out comes in sample space}}$$

The total number of outcomes that can occur = 4

The events which we require, is one

So,

Hence P(Blue) = 1/4 .... Ans

Example 4: A Dice is drawn. What is the probability of getting an even number?

Solution:

P of event =

$$\frac{\text{favorable number of out comes}}{\text{Total number of out comes in sample space}}$$

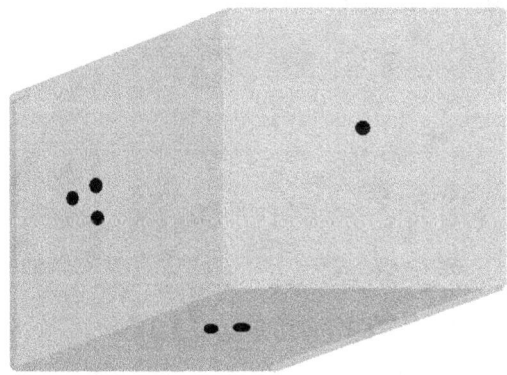

The total number of outcomes that can occur = 6

The events which we require, is 2, 4, or 6 = 3

So,

Hence P(Even number) = 3/6 = ½ .... Ans

Example 5: A Dice is drawn. What is the probability of getting an odd number?

Solution:

P of event =

$$\frac{\text{favorable number of out comes}}{\text{Total number of out comes in sample space}}$$

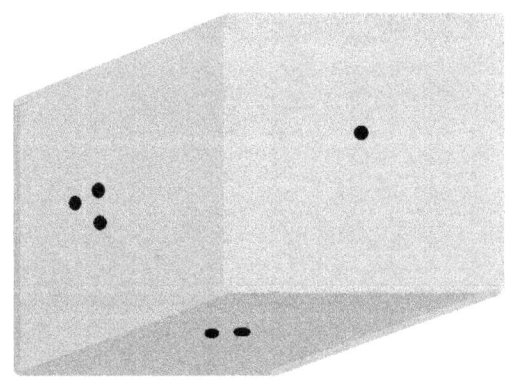

The total number of outcomes that can occur = 6

The events which we favor, is 1, 3, or 5 = 3

So,

Hence P(odd number) = 3/6 = ½   ……………. Ans

Example 6: I asked peter to choose any number between 1 to twenty. What is likely hood of Peter choosing 12?

Solution:

mathematics

P of event =

$$\frac{\text{required number}}{\text{Total number of out comes in sample space}}$$

The total number of outcomes that can occur = 20

The event of choosing 12 = 1

So,

Hence P(12) = 1/20   ……………. **Ans**

Example 7: I asked peter to choose any number between 1 to twenty. What is likely hood of Peter choosing any prime number?

Solution:

P of event =

$$\frac{\text{required number}}{\text{Total number of out comes in sample space}}$$

The total number of outcomes that can occur = 20

The event of choosing prime number

= 2, 3, 5, 7, 11, 13, 17, 19 = 8

So,

Hence P(Prime number) = 8/20 = 2/5   .... **Ans**

Example 8: There are 7 four-point stars 7 six point stars, 6 diamonds and 6 fivepoint stars in inside a box. If one thing is to be picked up with closed eyes from the box what is probability of a diamond to come up?

Solution:

mathematics

P of event =

$$\frac{\text{favorable out comes}}{\text{Total number of out comes in sample space}}$$

The total number of outcomes that can occur = 7 + 7 + 6 + 6 = 26

The event of choosing Blue ball = 7

So,

Hence P(diamond) = 6/26 .... **Ans**

Example 9: John tosses a coin 50 times. What is probability of getting heads.

The total number of outcomes that can occur = 50

The event of choosing heads = 25

So,

Hence P(Blue ball) = 25/50 = ½

This means out 50 john is likely to toss heads 25 times.

.... **Ans**

Example 10: Bonne is asked to pick a card from a pack of cards. What are the chances he picks a King.

Solution:

Pack of cards has 52 cards.

There are 4 kings in it.

Therefore chances are 4/ 52 = 1/ 13 .... **Ans**

Terminology

mathematics

Experiment

An experiment is a planned operation carried out under controlled conditions. Flipping coin twice or thrice is an example of an experiment.

Outcome:

A result of an experiment is called an outcome. The sample space of an experiment is the set of all possible outcomes. Three ways to represent a sample space are: to list the possible outcomes, to create a tree diagram, or to create a Venn diagram. The uppercase letter S is used to denote the sample space. For example, if you flip one fair coin, S = {H, T} where H = heads and T = tails are the outcomes.

Sample space

The sample space of an experiment is the set of all possible outcomes. Three ways to represent a sample space are: to list the possible outcomes, to create a tree diagram, or to create a Venn diagram. The uppercase letter S is used to denote the sample space. For example, if you flip one fair coin, S = {H, T} where H = heads and T = tails are the outcomes.

"Equally Likely"

In dice problem, there are six faces and all the faces have equal probability to show up. The out come of 1, 2, 3, 4, 5 or 6 may occur with equal probability. So it is called "Equally Likely" Equally likely means that each outcome of an experiment occurs with equal probability.

For example, if you toss a fair, six-sided die, each face (1, 2, 3, 4, 5, or 6) is as likely to occur as any other face. If you toss a fair coin, a Head (H) and a Tail (T) are equally likely to occur.

If you guess the answer to a true or false question in an exam, you are equally likely to select a correct answer or an incorrect answer.

Compliment event

The complement of event A is denoted A'. A' consists of all outcomes that are NOT in A.

Notice that $P(A) + P(A') = 1$.

For example, let S = {1, 2, 3, 4, 5, 6} and

let 4 coming up event is A = {4}.

Then, A'={1, 2, 3, 5, 6}.

$P(A) = 1/6$ and $P(A') = 5/6$

$P(A) + P(A') = 1/6 + 5/6 = 1$

Mutually exclusive

If we can not make two event at the same time then they are mutually exclusive.

The chances from head and tail are mutually exclusive in coin-toss. That means when a coin is tossed It can not land on both head and tail.

mathematics

But In cards if we randomly pick one card from a pack, we can get King of Clubs. So picking a king and picking a clubs are not mutually exclusive. Both can occur in a single event.

Sure – Event

Sure event is that event whose probability is 1

Impossible – Event

Impossible event is that event whose probability is 0

Example 11: Bonne is asked to pick a card from a pack of cards. What are the chances he picks a King.

Solution:

Pack of cards has 52 cards.

There are 4 kings in it.

Therefore chance of picking King

P(King) = 4/ 52 = 1/ 13

P'(King) = $1 - \frac{4}{52} = \frac{48}{52} = \frac{12}{13}$

or

Probability of not picking a King is = $1 - \frac{1}{13} = \frac{12}{13}$ ...

.... Ans

Example 12: There are 3 blue and 2 red marbles in a bag. What is the probability of drawing a blue marble on the first and second draw?

Solution:

P(blue) in first draw = 3/5

P(blue) in second draw = 2/4 = ½ .... **Ans**

Example 13: The probability of event is always between what values?

Ans: between 0 and 1

Example 14: Events are not affected by previous Events are called ?

Ans: Independent events

Example 15: A number is chosen at random from 1 to 10. Find the probability of not selecting a multiple of 2 or a multiple of 3.

Ans: 3/10

Example 15: Find the probability of drawing a 8 of Spades

Ans: 1/ 52

Example 16: A number is chosen at random from 1 to 10. Find the probability of not selecting a multiple of 3.

Ans: 7/ 10

Example 17: What is the total number of possible outcomes when rolling a pair of dice?

Ans 6 x 6 = 36.

Example 18: Find the probability of drawing a Heart from a pack of cards.

Ans: 1/ 4

Example 19: Find the probability of drawing cards K, Q, J, A.

Ans: 16/52 = 4/13

Example 20: Find the probability of drawing a black card.

Ans: ½

Example 21: Find the probability of rolling an odd prime number

Ans: 1/3

Example 22: When two dice are rolled, find the probability of rolling prime numbers on both dice.

Ans: ¼

Example23: Find the probability of drawing a Diamond. In pack of cards.

Ans ¼

Example 24: Find the probability of drawing a 8 of Clubs

Ans: 1/52

Example 25: List all possible outcomes from rolling a die.

Ans: 1 or 2 or 3 or 4 or 5 or 6. (Sample space = 6)

Example 26: Find the probability of drawing red cards 5 through 9

Ans: $\frac{5 \times 2}{52}$ = 5/26

Example 27: Find the probability of drawing a Joker when there is one joker in entire pack.

Ans: 1/53

Example 28: Find the probability of rolling a 4 or greater when you roll a dice or Find the probability of rolling at least 4 when you roll a dice

You can have 4, 5, 6

No of events/ sample space = 3/6

Ans: = ½

Example 29: Find the probability of not rolling a sum of 5

Solution:

We can have

Ans: 8/9

mathematics

Exercise

A fair coin is tossed.

1. Find all possible outcomes

Answer: _____

2. Find the probability of showing head

Answer: _____

3. Find the probability of showing tail

Answer: _____

4. Find the probability of showing either head or tail

Answer: _____

5. Find the probability of showing neither head nor tail

Answer: _____

Answers: 1) H, T 2) ½ 3) ½ 4) 1(always possible) 5) 0 (impossible)

That is all for the kids in this book.

Look for our books in amazon

By typing maths by M. saiprasad

www.ingramcontent.com/pod-product-compliance
Lightning Source LLC
Chambersburg PA
CBHW082341220526
45470CB00008B/2600